Name _____ Class ___

Skills Worksheet
Directed Reading

Section: Genetic Engineering
Complete each statement by underlining the correct term or phrase in the brackets.

1. Cohen and Boyer revolutionized genetics by producing recombinant [DNA / RNA].

2. In Cohen and Boyer's 1973 experiment, genetically engineered [bacterial / human] cells produced frog rRNA.

3. Moving genes from one organism to another is called [genetic / chemical] engineering.

4. [Restriction / Selection] enzymes recognize specific short sequences of DNA, then cut the DNA at specific sites within the sequences.

5. Circular DNA molecules that can replicate independently are called [plasmids / clones].

6. An enzyme called DNA [ligase / helicase] is used to help bond DNA fragments together.

7. In a genetic engineering experiment, making copies of a gene each time a host cell reproduces is called gene [cloning / reproduction].

8. Restriction enzyme cuts produce pieces of DNA with short single strands on each end that are called [sticky / recombinant] ends.

Study the following steps in a genetic engineering experiment. Determine the order in which the steps take place. Write the number of each step in the space provided.

_____ 9. The recombined vectors are returned to the host cell. The host cell reproduces.

_____ 10. DNA from the organism containing the gene of interest and the DNA from the vector are cut into pieces using restriction enzymes.

_____ 11. Cells that have received the gene of interest are identified.

_____ 12. The DNA fragments from the organism and the vector are combined using DNA ligase.

Name _____ Class _____ Date _____

Directed Reading *continued*

Complete each statement by writing the correct term or phrase in the space provided.

13. In a Southern blot, the DNA from each bacterial colony is isolated and cut into fragments by _____ _____ .

14. Gel _____ separates DNA fragments by their charge and size.

15. DNA fragments are _____ charged.

16. The larger a DNA fragment becomes, the _____ distance it travels in a gel.

17. After the DNA bands are separated, they are transferred to a piece of filter paper, which is moistened with a(n) _____ solution.

Name _____ Class _____ Date _____

Skills Worksheet

Directed Reading

Section: Human Applications of Genetic Engineering

Complete each statement by underlining the correct term or phrase in the brackets.

1. The research effort to determine the nucleotide sequence of the entire human genome and to map the location of every gene on each chromosome is called [the Human Genome Project / Project 2003].

2. Humans have about [30,000 / 120,000] genes.

3. Humans have about [3 billion / 2 trillion] base-pairs in all their DNA.

Complete each statement by writing the correct term or phrase in the space provided.

4. The protein _____ is produced by genetic engineering to treat diabetes, and the protein factor VIII is produced to treat _____ .

5. A(n) _____ is a solution containing a modified or killed version of a pathogen.

6. When a vaccine is injected, the immune system recognizes the pathogen's surface _____ and responds by making defensive proteins called _____ .

Read each question, and write your answer in the space provided.

7. What are two disadvantages of obtaining from living organisms the proteins needed to treat disease?

_____

_____

Name _____ Class _____ Date _____

Directed Reading continued

8. Why would a genetically engineered genital herpes vaccine be safer than one made from a naturally occurring herpes II virus?

_____
_____
_____
_____
_____
_____

9. What is the disadvantage of obtaining from living organisms the protein factor VIII?

_____
_____

10. What is a DNA fingerprint?

_____
_____

11. How is a DNA fingerprint made?

_____
_____
_____
_____

12. Explain why DNA fingerprints can be used to identify a person.

_____
_____
_____
_____

Name _____ Class _____ Date _____

Skills Worksheet
Directed Reading

Section: Genetic Engineering in Agriculture

Complete each statement by writing the correct term or phrase in the space provided.

1. Genetic engineers can add favorable characteristics to a plant by manipulating the plant's _____.

2. Some genetically engineered plants are now resistant to a(n) _____ called glyphosate.

3. Crops that are resistant to insects do not need to be sprayed by _____.

4. Some scientists are concerned that using glyphosate with GM crops could lead to glyphosate-resistant _____.

Read each question, and write your answer in the space provided.

5. Why does genetic technology make it easier to give cows growth hormones?

_____
_____
_____

6. Why are human genes added to the genes of farm animals?

_____
_____

7. What are transgenic animals?

_____
_____

8. What steps did Ian Wilmut take to clone Dolly?

_____
_____
_____
_____
_____

Copyright © by Holt, Rinehart and Winston. All rights reserved.

Holt Biology — Gene Technology

Name _____ Class _____ Date _____

Skills Worksheet

Active Reading

Section: Genetic Engineering

Read the passage below. Then answer the questions that follow.

Genetic engineering experiments use different approaches, but most share four basic steps.

Step 1: Cutting DNA. The DNA from the organism containing the gene of interest and the DNA from a vector are cut. The DNA is cut into pieces by restriction enzymes. **Restriction enzymes** are bacterial enzymes that recognize and bind to specific short sequences of DNA and then cut the DNA between specific nucleotides within the sequences. A **vector** is an agent that is used to carry the gene of interest into another cell. Commonly used vectors include viruses, yeast, and plasmids. **Plasmids** are circular DNA molecules that can replicate independently of the main chromosome of the bacteria. Plasmids are usually found in bacteria.

Step 2: Making Recombinant DNA. The DNA fragments from the organism containing the gene of interest are combined with the DNA fragments from the vector. An enzyme called DNA ligase is added to help bond the DNA fragments together.

Step 3: Cloning. In a process called **gene cloning,** many copies of the gene of interest are made each time the host cell reproduces. Since bacteria reproduce by binary fission, when a bacterial cell replicates its DNA, it also replicates its plasmid DNA.

Step 4: Screening. Cells that have received the particular gene of interest are distinguished from the cells that did not take up the vector with the gene of interest. Each time the cells reproduce, they make a copy of the gene of interest. The cells can transcribe and translate the gene to make the protein coded for in the gene.

SKILL: READING EFFECTIVELY

Read each question, and write your answer in the space provided.

1. What are restriction enzymes?

_____

_____

2. What is a vector?

_____

_____

Copyright © by Holt, Rinehart and Winston. All rights reserved.

Holt Biology   Gene Technology

Name _____ Class _____ Date _____

Active Reading *continued*

3. How are vectors and plasmids related?

_____

_____

4. What is the function of DNA ligase in genetic engineering?

_____

_____

5. What is the product of gene cloning?

_____

_____

6. How does the reproductive method of bacteria ensure the replication of plasmid DNA?

_____

_____

_____

_____

In the space provided, write the letter of the term or phrase that best completes the statement.

_____ **7.** All of the following are commonly used as vectors EXCEPT
  a. yeast.
  b. plasmids.
  c. DNA ligase.
  d. viruses.

Name _____ Class _____ Date _____

Skills Worksheet

Active Reading

Section: Human Applications of Genetic Engineering
Read the passage below. Then answer the questions that follow.

Many viral diseases, such as smallpox and polio, cannot be treated effectively by existing drugs. Instead they are combated by prevention, using vaccines. A **vaccine** is a solution containing a harmless version of a pathogen (disease-causing microorganism). When a vaccine is injected, the immune system recognizes the pathogen's surface proteins and responds by making defensive proteins called antibodies. In the future, if the same pathogen enters the body, the antibodies are there to combat the pathogen and stop its growth before it can cause disease.

Traditionally, vaccines have been prepared either by killing a specific pathogenic microbe or by making the microbe unable to grow. This ensures that the vaccine itself will not cause the disease. The problem with this approach is that there is a small but real danger that a failure in the process to kill or weaken the pathogen will result in transmission of the disease to the very patients seeking protection. This danger is one of the reasons why rabies vaccines are administered only when a person has actually been bitten by an animal suspected of carrying rabies.

SKILL: READING EFFECTIVELY
Read each question, and write your answer in the space provided.

1. What two viral diseases are identified in the first sentence of the passage?

   _____

2. What is a vaccine?

   _____

   _____

3. Why are the words enclosed by parentheses in the third sentence?

   _____

   _____

Copyright © by Holt, Rinehart and Winston. All rights reserved.

Holt Biology — Gene Technology

Name _____ Class _____ Date _____

Active Reading continued

4. One action of a body's immune system is to recognize a pathogen's surface proteins. What causes this action?

_____

_____

5. What is the effect of the immune system's action identified in question 4?

_____

_____

6. What two actions are taken to ensure that a vaccine will not cause disease?

_____

_____

_____

In the space provided, write the letter of the term or phrase that best completes the statement.

_____ **7.** Rabies vaccines are administered only when a person has been bitten by an animal suspected of carrying rabies because it is possible that the
   a. person receiving the vaccine may be allergic to it.
   b. pathogen may not have been killed or sufficiently weakened.
   c. animal may have a viral disease.
   d. Both (a) and (b)

Skills Worksheet
Active Reading

Section: Genetic Engineering in Agriculture

Read the passage below. Then answer the questions that follow.

In 1997, a scientist named Ian Wilmut captured worldwide attention when he announced the first successful cloning using differentiated cells from an adult animal. A differentiated cell is a cell that has specialized to become a specific type of cell (such as a liver cell or an udder cell). A lamb was cloned from the nucleus of an udder (mammary) cell taken from an adult sheep. Previously, scientists thought that cloning was only possible using embryonic or fetal cells that have not yet differentiated. Scientists thought that differentiated cells could not give rise to an entire organism. Wilmut's experiment proved otherwise.

An electric shock was used to fuse mammary cells from one sheep with egg cells without nuclei from a different sheep. The fused cells divided to form embryos, which were implanted into surrogate mothers. Only one embryo survived. Dolly, born on July 5, 1996, was genetically identical to the sheep that provided the mammary cell.

SKILL: READING EFFECTIVELY

Read each question, and write your answer in the space provided.

1. What announcement did Ian Wilmut make in 1997?

   _____

   _____

2. What is a differentiated cell?

   _____

   _____

3. Why is the word *mammary* enclosed in parentheses in the third sentence?

   _____

   _____

Name _____ Class _____ Date _____

Active Reading continued

4. What notions regarding cloning did Wilmut's experiment disprove?

_____

_____

_____

_____

SKILL: SEQUENCING INFORMATION

Study the following steps of Wilmut's experiment. Determine the order in which the steps took place. Write the number of each step in the space provided.

_____ **5.** Fused cells divided to form embryos.

_____ **6.** Mammary cells and egg cells were taken from two sheep.

_____ **7.** Dolly was born.

_____ **8.** Embryos were implanted into surrogate mothers.

_____ **9.** Electric shock was used to fuse cells.

In the space provided, write the letter of the term or phrase that best completes the statement.

_____ **10.** Dolly was genetically identical to the sheep that
   a. provided the mammary cell.
   b. was the surrogate mother.
   c. provided the egg cells.
   d. Both (a) and (b)

Copyright © by Holt, Rinehart and Winston. All rights reserved.

Holt Biology — Gene Technology

Name _____ Class _____ Date _____

Skills Worksheet

Vocabulary Review

Unscramble each listed term, and write the correct term in the space at right. In the space at left, write the letter of the description below that best matches the term or phrase.

_____ 1. nmauH nmeeGo tjPceor _____

_____ 2. tveorc _____

_____ 3. smipdal _____

_____ 4. brtnneacimo DAN _____

_____ 5. ccveina _____

_____ 6. ttiicrneosr meenyzs _____

_____ 7. gtnrcnaeis aailmn _____

_____ 8. neeg gninolc _____

_____ 9. ssierhpocrteelo _____

_____ 10. raeicngsnt aaimnl _____

_____ 11. eegncit ggeeiinnren _____

a. animal that has foreign DNA in its cells

b. a solution containing a weakened or modified version of a pathogen

c. a research effort to determine the nucleotide sequence of the human genome and map the location of every gene

d. an animal that has foreign DNA in its cells

e. bacterial enzymes that recognize and bind to specific short sequences of DNA, then cut the DNA at specific sites within the sequences

f. a technique that uses an electrical field within a gel to separate molecules by their size and charge

g. a circular DNA molecule that can replicate independently of the main chromosomes of bacteria

h. an agent that is used to carry the gene of interest into another cell

i. DNA made from two or more different organisms

j. when copies of the gene of interest are made each time the host cell reproduces

k. the process of manipulating genes for practical purposes

Name _____ Class _____ Date _____

Skills Worksheet

Science Skills

Interpreting Diagrams

Most genetic engineering experiments include the basic steps shown in the figure below. In this example, the gene responsible for producing insulin is the gene of interest. Use the figure to answer questions 1–4.

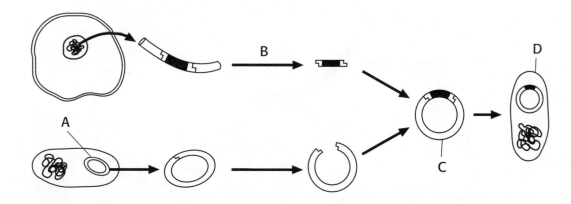

Read each question, and write your answer in the space provided.

1. Identify the structure labeled *A*. What is the function of this structure?

   _____

   _____

   _____

2. Explain the step labeled in *B*.

   _____

   _____

   _____

3. Identify the structure labeled *C*.

   _____

   _____

4. Identify the structure labeled *D*.

   _____

   _____

Science Skills continued

Restriction enzymes recognize specific short nucleotide sequences in DNA and cut them within those sequences, resulting in single stranded areas called sticky ends, as shown in the figure below. In order for the plasmid DNA to recombine successfully with the gene of interest, the sticky ends of the plasmid DNA must pair with the complementary sticky ends of the gene of interest. Use the figure to answer questions 5–7.

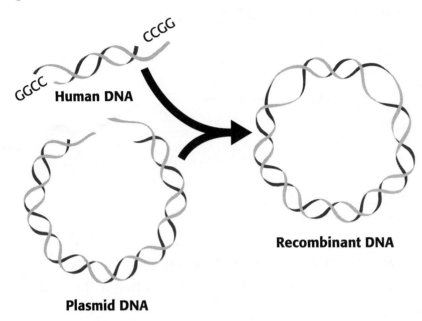

Read each question, and write your answer in the space provided.

5. List the two nucleotide sequences that are complementary to the sticky end sequences on the human DNA.

   _____

6. List the paired nucleotide sequences of the recombinant DNA.

   _____

7. The gene for tetracycline resistance is present in the plasmid DNA. Explain the reason for using plasmid DNA that contains the gene for tetracycline resistance.

   _____
   _____
   _____
   _____
   _____

Name _____ Class _____ Date _____

Skills Worksheet
Concept Mapping

Using the terms and phrases provided below, complete the concept map showing the uses and applications of gene technology.

agriculture   genetic disorders   probes
cloned animals   genetic engineering   restriction enzymes
electrophoresis   medicines   vaccines

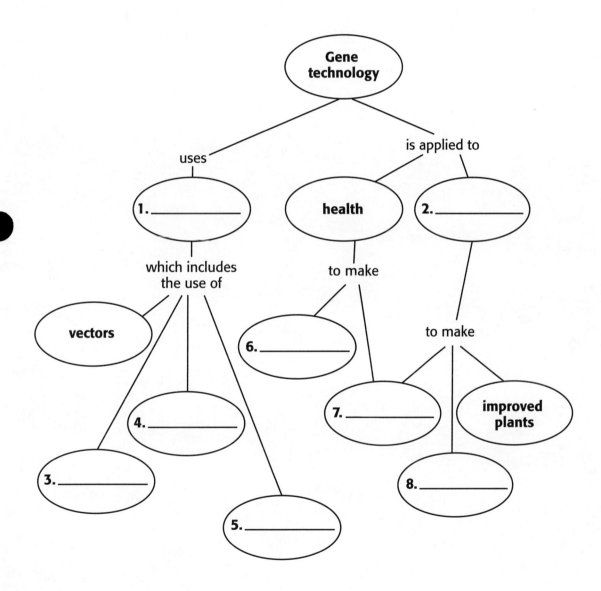

Holt Biology   Gene Technology

Name _____ Class _____ Date _____

Skills Worksheet

Critical Thinking

Work-Alikes

In the space provided, write the letter of the term or phrase that best describes how each numbered item functions.

_____ 1. cloning

_____ 2. vector

_____ 3. Human Genome Project

_____ 4. cutting DNA plasmid

a. delivery service

b. photocopying

c. removing a piece from a doughnut

d. making a map

Cause and Effect

In the space provided, write the letter of the term or phrase that best matches each cause or effect given below.

Cause	Effect
5. _____	separation of DNA fragments
6. restriction enzymes	_____
7. _____	DNA fragment bonds to DNA vector
8. _____	recombinant DNA
9. gene of interest is isolated	_____

a. DNA ligase is added

b. it can be used to determine its sequence of nucleotides

c. gel electrophoresis

d. genetic engineering

e. DNA fragments with single-stranded "sticky ends"

Trade-offs

In the space provided, write the letter of the bad news item that best matches each numbered good news item below.

Good News

_____ 10. Recombinant DNA from some pathogens produces harmless vaccines.

_____ 11. Factor VIII is produced by genetic engineering.

_____ 12. Glyphosate-resistant crops have been genetically engineered.

_____ 13. Insulin can be genetically engineered.

Bad News

a. The use of glyphosate may lead to glyphosate-resistant weeds.

b. Some traditional vaccines can cause the same disease in patients.

c. Diabetes patients cannot make adequate amounts of this hormone.

d. Hemophiliacs require blood factors for blood clotting.

Copyright © by Holt, Rinehart and Winston. All rights reserved.

Holt Biology — Gene Technology

Critical Thinking continued

Linkages

In the spaces provided, write the letters of the two terms or phrases that are linked together by the term or phrase in the middle. The choices can be placed in any order.

14. _____ production of recombinant DNA _____

15. _____ resistance gene in bacteria in some recombinant DNA _____

16. _____ bacteria-produced growth hormone _____

17. _____ milk containing human proteins _____

18. _____ embryos formed and implanted into surrogate mother _____

a. cells with resistance gene survive
b. DNA directs production of large quantities of protein
c. sheep mammary cells fused to eggs without nuclei
d. DNA cut into pieces
e. growth hormone genes
f. proteins extracted from milk
g. increased milk production
h. transgenic animal
i. tetracycline added to culture of recombined DNA
j. cloned sheep

Analogies

An analogy is a relationship between two pairs of terms or phrases written as a : b :: c : d. The symbol : is read as "is to," and the symbol :: is read as "as." In the space provided, write the letter of the pair of terms or phrases that best completes the analogy shown.

_____ 19. vector : DNA fragment ::
   a. DNA fingerprint : human fingerprint
   b. anticoagulant : gene
   c. RFLP : clone
   d. plasmid : gene

_____ 20. restriction enzyme : cutting DNA ::
   a. cleaving DNA : identifying DNA
   b. cloning DNA : radioactively labeling DNA
   c. screening bacterial cell : cloning DNA
   d. gel electrophoresis : separating DNA

Critical Thinking continued

_____ 21. cow growth hormone gene : bacteria ::
   a. weedkiller-resistance gene : crops
   b. glyphosate gene : cows
   c. nitrogen-fixing gene : hogs
   d. pest-resistance gene : insects

_____ 22. plasmid : vector ::
   a. Eco R1 : frog ribosomal RNA gene
   b. factor VIII : blood-clotting protein
   c. cloning : screening
   d. PCR : probe

Name _____ Class _____ Date _____

Skills Worksheet

Test Prep Pretest

In the space provided, write the letter of the term or phrase that best completes each statement or best answers each question.

_____ 1. How are genetically engineered vaccines different from those made from weakened pathogens?
   a. They are ineffective.
   b. They cause only a mild form of the disease.
   c. They eliminate the risk of transmitting the disease to the person injected.
   d. They cause the immune system to make antibodies.

_____ 2. Tetracycline is used in genetic engineering experiments as a way to
   a. identify bacteria that have taken up the recombined plasmid.
   b. produce stronger strains of bacteria.
   c. prevent the cultures from becoming infected with bacteria.
   d. kill cell clones that contain recombinant DNA.

_____ 3. Which genes of a pathogen are used to make a genetically engineered vaccine?
   a. those that encode a pathogen's surface proteins
   b. those that are harmless
   c. those that are similar to human genes
   d. those that encode antibodies

_____ 4. Gene technology is used to improve farm animals in which of the following ways?
   a. increasing milk production using cow growth hormone produced by bacteria
   b. cloning herds of animals to make medically useful human proteins
   c. producing milk containing human proteins by adding human genes to farm-animal genes
   d. All of the above

Name _____ Class _____ Date _____

Test Prep Pretest *continued*

Questions 5–7 refer to the figure below, which shows the steps of a genetic engineering experiment using DNA from a human insulin gene.

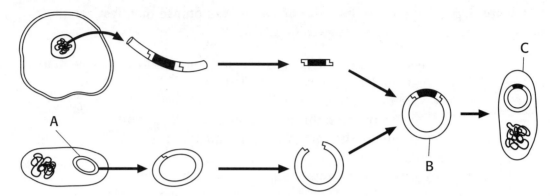

_____ 5. The structure labeled *A* is called
  a. plasmid DNA.
  b. a vector.
  c. a restriction enzyme.
  d. Both (a) and (b)

_____ 6. In *B*, the DNA of the gene and the vector are
  a. cloned.
  b. isolated.
  c. recombined.
  d. cut by the restriction enzyme.

_____ 7. In *C*, the
  a. gene is cloned.
  b. cells are screened.
  c. recombined plasmid DNA is inserted into the bacterium.
  d. DNA is cut.

Complete each statement by writing the correct term or phrase in the space provided.

8. The first step of Cohen and Boyer's genetic engineering experiment was to isolate the _____ of interest from the DNA of an African clawed frog.

9. Recombinant DNA is made when a DNA fragment is put into the DNA of a(n) _____ .

10. Any two fragments of DNA cut by the same restriction enzyme can pair because their ends are _____ .

11. Genetic engineering has benefited humans afflicted with diabetes by developing bacteria that produce _____ .

Name _____ Class _____ Date _____

Test Prep Pretest *continued*

12. By using the genetically engineered blood-clotting agent _____ _____ , hemophiliacs can eliminate the risks associated with blood products obtained from other individuals.

13. A vaccine is a solution that contains all or part of a harmless version of a(n) _____ .

14. Crop plants that are resistant to the biodegradable weedkiller _____ have been developed.

15. In genetic engineering, the enzyme _____ _____ helps the DNA fragments bind.

Read each question, and write your answer in the space provided.

16. Describe the Human Genome Project.

_____

_____

_____

_____

17. List two ways in which DNA fingerprints are used.

_____

_____

_____

18. Explain why the development of genetically engineered proteins has been important to pharmaceutical companies.

_____

_____

_____

_____

_____

_____

Test Prep Pretest continued

19. Why is the development of plants that are resistant to insects important?

_____

_____

_____

20. Explain the first step of Ian Wilmut's successful cloning experiment.

_____

_____

_____

_____

_____

_____

_____

Name _____ Class _____ Date _____

[Assessment]

Quiz

Section: Genetic Engineering

In the space provided, write the letter of the description that best matches the term or phrase.

_____ 1. genetic engineering

_____ 2. vector

_____ 3. plasmid

_____ 4. cloning

_____ 5. recombinant DNA

_____ 6. restriction enzymes

_____ 7. gel electrophoresis

a. made from DNA from two separate organisms

b. uses an electrical field to separate molecules

c. growing a large number of genetically identical cells from a single cell

d. isolates a gene from one organism's DNA and recombines it with another organism's DNA

e. can carry a DNA fragment into another cell

f. can be used as a vector

g. used to cut DNA at specific sequences

In the space provided, write the letter of the term or phrase that best completes each statement or best answers each question.

_____ 8. Radioactive or fluorescent-labeled RNA or single-stranded DNA pieces that are complementary to the gene of interest and are used to confirm the presence of a cloned gene are called
   a. probes.
   b. plasmids.
   c. vaccines.
   d. clones.

_____ 9. Plasmids
   a. are circular pieces of bacterial DNA.
   b. can replicate independently of the organism's main chromosome.
   c. are often used as vectors in genetic engineering experiments.
   d. All of the above

_____ 10. Genetic engineering experiments use tetracycline to
   a. prevent the cell cultures from becoming infected.
   b. kill the cells that had the recombinant DNA.
   c. identify cells that have taken up the recombined vector DNA.
   d. All of the above

Copyright © by Holt, Rinehart and Winston. All rights reserved.

Holt Biology — Gene Technology

Name _____ Class _____ Date _____

Assessment

Quiz

Section: Human Applications of Genetic Engineering

In the space provided, write the letter of the description that best matches the term or phrase.

_____ 1. DNA fingerprinting

_____ 2. Human Genome Project

_____ 3. exon

_____ 4. human genome

_____ 5. vaccine

_____ 6. forensics

a. includes only nucleotides that are transcribed and translated

b. contains only about 30,000–40,000 genes

c. made from a harmless version of a disease-causing agent

d. investigates the cause of injury or death in crimes

e. identified base pairs in the human cells

f. made from DNA fragments that are separated and probed

In the space provided, write the letter of the term or phrase that best completes each statement or best answers each question.

_____ 7. Taking genetically engineered drugs helps people with hemophilia avoid
   a. contracting HIV.
   b. contracting hepatitis.
   c. prolonged bleeding.
   d. All of the above

_____ 8. People with juvenile diabetes can live with their disease with the help of genetically engineered
   a. factor VIII.
   b. insulin.
   c. taxol.
   d. growth factors.

_____ 9. To make a DNA fingerprint, which of the following is NOT required?
   a. forensics
   b. gel electrophoresis
   c. radioactive probes
   d. X-ray film

_____ 10. To make a genetically engineered vaccine, the genes that encode a pathogen's surface proteins are inserted into
   a. humans.
   b. another pathogen.
   c. a harmless virus.
   d. antibodies.

Name _____ Class _____ Date _____

Assessment

Quiz

Section: Genetic Engineering in Agriculture

In the space provided, write the letter of the description that best matches the term or phrase.

_____ 1. transgenic animal

_____ 2. differentiated cell

_____ 3. glyphosate

_____ 4. Dolly

_____ 5. embryonic cell

a. weedkiller that some crop plants are genetically altered to be resistant to

b. first successfully cloned animal using differentiated cells

c. is programmed to become a specific type of cell

d. an animal with foreign DNA in its cells

e. is not yet programmed to become a specific type of cell

In the space provided, write the letter of the term or phrase that best completes each statement or best answers each question.

_____ 6. Genetic engineering has been used to
   a. improve crop plants.
   b. modify farm animals.
   c. clone animals.
   d. All of the above

_____ 7. Some scientists are concerned that modifying crops using genetic engineering could result in
   a. weedkiller-resistant weeds.
   b. widespread food allergies.
   c. less nutritious foods.
   d. gene flow across plant species.

_____ 8. Crop plants have been modified by genetic engineering to make them
   a. more nutritious.
   b. resistant to insects.
   c. resistant to weedkillers.
   d. All of the above

_____ 9. In which organism have genes for cow growth hormones been inserted?
   a. pigs
   b. humans
   c. bacteria
   d. algae

_____ 10. Farm animals have been made to produce human proteins in their milk by
   a. giving them more nutritious GM feed.
   b. adding human genes to their DNA.
   c. adding growth hormone to their diet.
   d. injecting them with transgenic bacteria.

Copyright © by Holt, Rinehart and Winston. All rights reserved.
Holt Biology   Gene Technology

Name _____ Class _____ Date _____

Assessment

Chapter Test

Gene Technology

In the space provided, write the letter of the description that best matches the term or phrase.

_____ 1. genetic engineering

_____ 2. vector

_____ 3. plasmid

_____ 4. cloning

_____ 5. recombinant DNA

_____ 6. restriction enzymes

_____ 7. DNA fingerprinting

_____ 8. Southern blot

_____ 9. gel electrophoresis

_____ 10. transgenic animal

_____ 11. differentiated cell

a. an animal with foreign DNA in its cells

b. made from DNA from two separate organisms

c. is programmed to become a specific type of cell

d. uses an electrical field to separate molecules

e. growing a large number of genetically identical cells from a single cell

f. isolates a gene from one organism's DNA and recombines it with another organism's DNA

g. pattern of dark bands on X-ray film made when DNA fragments are separated and probed

h. uses radioactively labeled RNA or single-stranded DNA as a probe to identify specific genes

i. can carry a DNA fragment into another cell

j. can be used as a vector

k. used to cut DNA at specific sequences

In the space provided, write the letter of the term or phrase that best completes each statement or best answers each question.

_____ 12. Scientists can now genetically engineer vaccines by inserting the genes that encode the pathogen's surface proteins into
   a. a specific pathogenic microbe.
   b. the DNA of harmless bacteria or viruses.
   c. the DNA of the patient.
   d. the RNA of the defective genes.

_____ 13. Which of the following is a potential benefit of genetically altered crops?
   a. Pesticide use is reduced.
   b. Tolerance of environmental stress is increased.
   c. Food spoilage is reduced.
   d. All of the above

Name _____ Class _____ Date _____

Chapter Test *continued*

_____ 14. Which of the following is NOT an example of gene technology used in farming?
   a. the use of cow growth hormone produced by bacteria to increase milk production in cows
   b. the development of larger and faster-growing breeds of livestock
   c. the cloning of human brain cells from selected farm animals
   d. the addition of human genes to farm-animal genes to obtain milk containing human proteins

_____ 15. Ian Wilmut's cloning of the sheep Dolly in 1997 was considered a breakthrough in genetic engineering because
   a. scientists thought cloning was impossible.
   b. scientists thought only fetal cells could be used to produce clones.
   c. scientists had never before isolated mammary cells.
   d. sheep had never responded well to gene technology procedures.

_____ 16. A cell that has specialized and become a specific type of cell is called a
   a. clone.
   b. plasmid.
   c. transgenic cell.
   d. differentiated cell.

_____ 17. A gene that codes for resistance to glyphosate has been added to the genome of certain plants. These plants will
   a. produce chemicals that kill weeds growing near them.
   b. die when exposed to glyphosate.
   c. survive when glyphosate is applied to the field.
   d. convert glyphosate into fertilizer.

_____ 18. The Human Genome Project has brought scientists from around the world together in order to
   a. identify the base pair sequence of all human genes.
   b. reduce the number of base pairs needed to code for human genes.
   c. eliminate the introns in human DNA.
   d. All of the above

_____ 19. Taking genetically engineered drugs helps people with hemophilia avoid
   a. the need for insulin.
   b. viruses from blood transfusions.
   c. contracting diabetes.
   d. the need for factor VIII.

_____ 20. DNA fingerprints can be used to
   a. identify genes that cause genetic disorders.
   b. establish whether two people are related.
   c. solve violent crimes.
   d. All of the above

Name _____ Class _____ Date _____

Assessment

Chapter Test

Gene Technology

In the space provided, write the letter of the term or phrase that best completes each statement or best answers each question.

_____ 1. Recombinant DNA is formed by joining DNA molecules
   a. from two different species.
   b. with a carbohydrate from a different species.
   c. with RNA molecules.
   d. with a protein from a different species.

_____ 2. Fragments of DNA that have complementary sticky ends
   a. are found only in bacterial cells.
   b. can join with each other.
   c. can only join with complementary fragments from the same species.
   d. are immediately digested by enzymes in the cytoplasm of the cell.

_____ 3. The risk associated with vaccines prepared from dead or weakened pathogenic microbes is that
   a. a few remaining live or unweakened microbes could still cause the disease.
   b. the antibodies that result may not work.
   c. the vaccine protects only against other diseases.
   d. None of the above

_____ 4. Which of the following does NOT describe a probe?
   a. single strand of nucleotides
   b. complementary to the gene of interest
   c. short pieces of artificial DNA used to make copies of genes
   d. labeled with fluorescent or radioactive substances

_____ 5. What role does electrophoresis play in identifying a specific gene?
   a. It cuts the gene of interest out of DNA at specific points.
   b. It kills all cells that have not taken up the gene of interest.
   c. It binds the gene of interest with probes.
   d. It separates DNA fragments by size.

_____ 6. Which of the following was a surprise to scientists working on the Human Genome Project?
   a. There were fewer genes than they had predicted.
   b. There were many more exons than they had predicted.
   c. There were many more base pairs than they had predicted.
   d. The DNA was much longer than they had predicted.

Copyright © by Holt, Rinehart and Winston. All rights reserved.

Holt Biology • Gene Technology

Name _____ Class _____ Date _____

Chapter Test continued

_____ 7. Genetically engineered drugs help people with hemophilia by
  a. helping them fight hepatitis and AIDS infections.
  b. preventing their blood from clotting too much.
  c. repairing their genes so they can make their own factor VIII.
  d. providing blood clotting factors without the risk of AIDS infection.

_____ 8. Which of the following correctly pairs a genetically engineered medicine with its use?
  a. GM insulin is used to treat anemia.
  b. GM growth hormone is used to treat diabetes.
  c. GM interferons are used to treat viral infections.
  d. GM taxol is used to treat burns and ulcers.

_____ 9. Which of the following is NOT a use of DNA fingerprints?
  a. distinguishing whether a clone arose from a differentiated cell
  b. establishing whether two people are related to each other
  c. identifying genes that cause specific genetic disorders
  d. determining whether suspects could have been involved in violent crimes

_____ 10. Genetic engineers are developing approaches for improving agriculture in all of the following ways EXCEPT
  a. making plants resistant to insects.
  b. making plants resistant to weedkillers.
  c. enabling plants to produce medically useful proteins.
  d. improving the nutritional value of certain plants.

Complete each statement by writing the correct term or phrase in the space provided.

11. The pattern of dark bands on X-ray film made when an individual's DNA fragments are separated, probed, and then exposed to an X-ray film is called a(n) _____ _____ .

12. Small, circular forms of bacterial DNA that can replicate independently of the main bacterial chromosome are called _____ .

13. Genetically identical cells grown from a single cell are called _____ .

14. Enzymes that cut DNA at specific sequences, producing fragments of DNA, are called _____ _____ .

Name _____ Class _____ Date _____

Chapter Test *continued*

15. Scientists use a technique called a(n) _____ _____ to confirm that the cloned gene of interest is present in a bacterial colony.

16. Radioactive or fluorescent-labeled RNA or single-stranded DNA pieces that are complementary to the gene of interest are called _____ .

17. Animals that contain foreign DNA are called _____ animals.

18. A cell that has been specialized to become a specific type of cell is called a(n) _____ cell.

19. Growing bacterial cells in the presence of _____ allows a researcher to identify which cells contain the gene of interest.

20. When cut with restriction enzymes, DNA fragments have short single-stranded ends that are _____ to each other. These ends are called _____ ends.

Read each question, and write your answer in the space provided.

21. Describe three ways genetic engineering has been used to improve plants.

_____
_____
_____
_____

22. Summarize two ways genetic engineering techniques have been used to modify farm animals.

_____
_____
_____
_____
_____

Name _____ Class _____ Date _____

Chapter Test continued

23. Summarize the four steps of a genetic engineering experiment.

24. Describe two different uses for DNA fingerprints.

25. Explain how gel electrophoresis is used in genetic engineering experiments.

Name _____ Class _____ Date _____

Quick Lab

DATASHEET FOR IN-TEXT LAB

Modeling Gel Electrophoresis

You can use beads to model how DNA fragments are separated in a gel during electrophoresis.

MATERIALS
- 500 mL beaker
- large jar
- 3 sets of beads—each set a different size and different color

Procedure

1. Fill a large jar with the largest beads. The filled jar represents a gel.
2. Mix the two smaller beads in the beaker and then pour them slowly on top of the "gel." The two smaller size beads represent DNA fragments of different sizes.
3. Observe the flow of the beads through the "gel." Lightly agitate the jar if the beads do not flow easily.

Analysis

1. **Identify** which beads flowed through the "gel" the fastest.

   _____

2. **Relate** the sizes of the beads to the sizes of DNA fragments.

   _____

   _____

3. **Determine** whether the top or the bottom of the jar represents the side of the gel with the positively charged pole.

   _____

   _____

   _____

4. **Critical Thinking**
   Forming Conclusions Why do the beads you identified in Analysis question 1 pass through the "gel" more quickly?

   _____

   _____

   _____

Name _____ Class _____ Date _____

Exploration Lab

DATASHEET FOR IN-TEXT LAB

Modeling Recombinant DNA

SKILLS
- Modeling
- Comparing

OBJECTIVES
- **Construct** a model that can be used to explore the process of genetic engineering.
- **Describe** how recombinant DNA is made.

MATERIALS
- paper clips (56)
- plastic soda straw pieces (56)
- pushpins (15 red, 15 green, 13 blue, and 13 yellow)

Before You Begin

Genetic engineering is the process of taking a gene from one organism and inserting it into the DNA of another organism. The gene is delivered by a **vector**, such as a virus, or a bacterial **plasmid**.

First, a fragment of a chromosome that contains the gene is isolated by using a **restriction enzyme**, which cuts DNA at a specific nucleotide-base sequence. Some restriction enzymes cut DNA unevenly, producing single-stranded **sticky ends.** The DNA of the vector is cut by the same restriction enzyme. Next, the chromosome fragment is mixed with the cut DNA of the vector. Finally, an enzyme called **DNA ligase** joins the ends of the two types of cut DNA, producing **recombinant DNA.**

In this lab, you will model genetic engineering techniques. You will simulate the making of recombinant DNA that has a human gene inserted into the DNA of a plasmid.

1. Write a definition for each boldface term in the paragraph above and for the term *base-pairing rules*. Use a separate sheet of paper.

2. Based on the objectives for this lab, write a question you would like to explore about the process of genetic engineering.

_____

_____

Copyright © by Holt, Rinehart and Winston. All rights reserved.

Holt Biology — Gene Technology

Name _____ Class _____ Date _____

Modeling Recombinant DNA *continued*

Procedure
PART A: MODEL GENETIC ENGINEERING

1. Make 56 model nucleotides. To make a nucleotide, insert a pushpin midway along the length of a 3 cm piece of a soda straw. **CAUTION: Handle pushpins carefully. Pointed objects can cause injury.** Push a paper clip into one end of the soda-straw piece until it touches the pushpin.

2. Begin a model of a bacterial plasmid by arranging nucleotides for one DNA strand in the following order: blue, red, green, yellow, red, red, blue, blue, green, red, blue, green, red, blue, blue, green, yellow, and red. Join two adjacent nucleotides by inserting the paper clip end of one into the open end of the other.

3. Using your first DNA strand and the base-pairing rules, build the complementary strand of plasmid DNA. **Note:** *Yellow is complementary to blue, and green is complementary to red.*

4. Complete your model of a circular plasmid by joining the opposite ends of each DNA strand. Make a sketch showing the sequence of bases in your model plasmid. Use the abbreviations B, Y, G, and R for the pushpin colors. Your sketch should be similar to the one shown below.

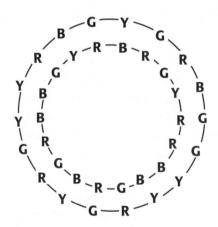

5. Begin a model of a human chromosome fragment made by a restriction enzyme. Place nucleotides for one DNA strand in the following order: BBRRYGGBRY. Build the second DNA strand by arranging the remaining nucleotides in the following order: BRRYGBYYGG.

6. Match the complementary portions of the two strands of DNA you made in step 5. Pair as many base pairs in a row as you can. Make a sketch showing the sequence of bases in your model of a human chromosome fragment.

7. Imagine that the restriction enzyme that cut the human chromosome fragment you made in steps 5 and 6 is moving around your model plasmid until it finds the sequence YRRBBG and its complementary sequence, BGGYYR. This restriction enzyme cuts each sequence between a B and a G. Find such a section in your sketch of your model plasmid's DNA.

Copyright © by Holt, Rinehart and Winston. All rights reserved.
Holt Biology 42 Gene Technology

Name _____ Class _____ Date _____

Modeling Recombinant DNA continued

8. Simulate the action of the restriction enzyme on the section you identified in step 7. Open both strands of your model plasmid's DNA by pulling apart the adjacent green and blue nucleotides in each strand. Make a sketch of the split plasmid DNA molecule.

9. Move your model human DNA fragment into the break in your model plasmid's DNA molecule. Imagine that a ligase joins the ends of the human and plasmid DNA. Make a sketch of your final model DNA molecule.

PART B: CLEANUP AND DISPOSAL

10. Dispose of damaged pushpins in the designated waste container.

11. Clean up your work area and all lab equipment. Return lab equipment to its proper place. Wash your hands thoroughly before you leave the lab and after you finish all work.

Analyze and Conclude

1. **Comparing Structures** Compare your models of plasmid DNA and human DNA.

_____

_____

_____

_____

_____

2. **Relating Concepts** What do the sections of four unpaired nucleotides in your model human DNA fragment represent?

_____

_____

3. **Comparing Structures** How did your original model plasmid DNA molecule differ from your final model DNA molecule?

_____

_____

_____

Copyright © by Holt, Rinehart and Winston. All rights reserved.

Name _____ Class _____ Date _____

Modeling Recombinant DNA *continued*

4. Drawing Conclusions What does the molecule you made in step 9 represent?

_____

_____

5. Further Inquiry Write a new question that could be explored with another investigation.

_____

_____

Name _____ Class _____ Date _____

Exploration Lab

BIOTECHNOLOGY

DNA Fingerprinting

With the exception of identical twins, no two people have the same DNA sequence. The slight differences (or polymorphisms) in each person's DNA sequence ensure that each person (except identical twins) has a unique set of sites where restriction enzymes can bind and cut the DNA. These differences are know as *restriction fragment length polymorphisms (RFLPs)*. Due to RFLPs, the exact number and size of fragments produced by digestion with a specific restriction enzyme varies from person to person. When the fragments are separated on a gel by electrophoresis, a unique banding pattern results that is known as a DNA fingerprint. An unknown DNA sample can be matched with a known sample of a person's DNA to an accuracy of 1 in 10 billion people. Due to its accuracy, DNA fingerprinting is often used in the legal system to determine identity in criminal cases and to establish hereditary relationships, such as paternity.

In this lab, you will perform some of the experimental procedures involved in DNA fingerprinting and use your results to identify a hypothetical burglar.

OBJECTIVES

Conduct a restriction digestion of DNA samples.

Use agarose gel electrophoresis to separate the digested DNA samples.

Evaluate the results of simulated DNA fingerprints.

Identify a hypothetical burglar by analyzing simulated DNA fingerprints.

MATERIALS

- agarose gel on gel tray (0.8%)
- beaker, 500 mL
- DNA samples (5)
- DNA stain
- electrophoresis system, battery powered
- gel staining tray
- gloves
- graduated cylinder, 250 mL
- lab apron
- Lambda DNA/*Hind III* marker
- loading dye
- microfuge tubes (6)
- micropipet, 10 or 20 µL
- micropipet tips (30)
- *Pvu II* restriction enzyme
- *Pvu II* reaction buffer, 10x
- microfuge tube rack/float
- TBE running buffer, 350 mL (1×)
- safety goggles
- water bath, 37° C
- water bath, 65° C

Copyright © by Holt, Rinehart and Winston. All rights reserved.

Holt Biology 45 Gene Technology

Name _____ Class _____ Date _____

DNA Fingerprinting *continued*

Procedure

1. Read the following scenario.

   The police are investigating a burglary. The police report indicates that as the police arrived, the masked burglar smashed a glass door with a chair and quickly escaped. Police found small pieces of bloodstained fabric on some of the pieces of the glass door. Four brothers have been brought in as suspects. Because all of them have the same blood type, investigators have resorted to DNA fingerprinting to identify which brother is the burglar. You have been provided with the DNA from the blood samples found at the crime scene and DNA from blood samples taken from each of the four suspects.

PART 1: RESTRICTION ENZYME DIGESTION OF DNA

In the first step of DNA fingerprinting, known and unknown samples are obtained and then digested, or cut into small fragments (RFLPs), by the same restriction enzyme.

2. Put on safety goggles and a lab apron.

3. Obtain five microcentrifuge tubes and label as follows: Crime Scene, Suspect #1, Suspect #2, Suspect #3, Suspect #4.

4. Set the micropipet to 2 μL and place a clean tip on the pipet.

5. Add 2 μL of *Pvu* II 10x reaction buffer to each tube.

6. Set the micropipet to 10 μL.

7. Using a fresh pipet tip for each sample, add 10 μL of each DNA sample to the appropriate tube.

8. Set the micropipet to 2 μL.

9. Add 2 μL of Pvu II restriction enzyme to each tube. Again, be sure to use a fresh micropipet tip for each sample. *Note: It may be helpful to place the pipet tip against the side of the microcentrifuge tube when dispensing small volumes.*

10. Close each tube and snap the tubes in a downward motion with your wrist to force all of the reagents to the bottom of the tubes.

11. Incubate all five tubes at 37° C for 45–60 minutes.

12. Add 2 μL of loading dye to each tube.

13. Incubate all five tubes for five minutes at 65° C to stop the restriction enzyme activity.

14. Place the samples in the freezer if you will not be using them immediately. Make sure all samples for your group are stored together and labeled with your group name.

Name _____ Class _____ Date _____

| DNA Fingerprinting *continued*

PART 2: GEL ELECTROPHORESIS OF DNA SAMPLES

The next step of a DNA fingerprint is to separate the RFLPs by gel electrophoresis. In gel electrophoresis, the digested DNA samples are loaded into wells on a jellylike slab called a gel. The gel is then exposed to an electrical current. DNA has a negative electrical charge, so the RFLPs are attracted to the positive pole when an electric current is applied. Smaller fragments travel farther through the gel than longer ones. A special dye, called loading dye, is added to the DNA samples for two reasons. The loading dye is heavier than the buffer, which helps the samples to stay in the wells. The loading dye also runs slightly faster than the DNA samples and indicates when the gel has finished running.

15. Place the tray with the gel on the lab bench.

16. Set the pipet to 10 μL and place a new tip on the end of your micropipet.

17. Open the microtube containing the DNA marker and use the pipet to load 10 μL of the marker DNA into the well in Lane 1 of an agarose gel. To do this, place both elbows on the lab table, lean over the gel, and slowly lower the micropipet tip into the opening of the well before depressing the plunger. *Note: Do not jab the micropipet tip through the bottom of the well.*

18. Set the pipet to 15 μL and place a new tip on the end of your micropipet.

19. Open the microtube containing the crime scene sample and use the pipet to load 15 μL of the sample into the well in Lane 2 of an agarose gel.

20. Using a new micropipet tip for each tube, repeat step 19 for each of the remaining samples. Place each sample in a well, according to the following lane assignments:

    Lane 1   DNA marker
    Lane 2   Crime scene DNA
    Lane 3   Suspect #1 DNA
    Lane 4   Suspect #2 DNA
    Lane 5   Suspect #3 DNA
    Lane 6   Suspect #4 DNA

21. Carefully place the agarose gel (still in a gel-casting tray) in the electrophoresis chamber of an electrophoresis apparatus, shown in **Figure 1,** so that the wells are closest to the negative electrode.

22. Pour approximately 350 mL of 1×TBE running buffer into a beaker.

23. *Gently* and *slowly* pour the running buffer from the beaker into one side of the electrophoresis chamber until the gel is completely covered (approximately 1 to 2 mm above the top surface of the gel). *Note: Pouring too fast will rinse your DNA sample out of the wells. Be careful not to overfill the chamber with buffer.*

DNA Fingerprinting *continued*

FIGURE 1 SETTING UP THE ELECTROPHORESIS APPARATUS

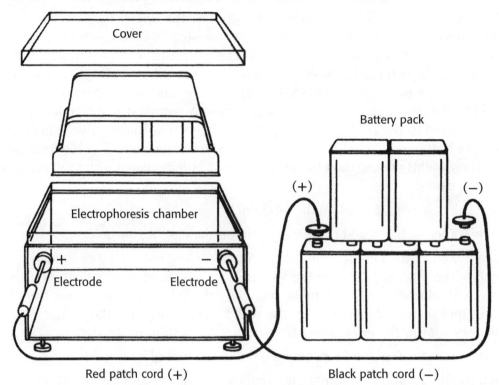

24. Place the cover on the electrophoresis chamber. Wipe off any spills around the electrophoretic apparatus before doing the next step.

25. Connect five 9 V alkaline batteries as shown in **Figure 1. CAUTION: Do not touch both ends of the patch cords or both terminals on the battery pack at the same time.**

26. Connect the red (positive) patch cord to the red terminal on the chamber and the red terminal on the battery pack. Follow the same procedure with the black (negative) patch cord and the black terminals.

27. Observe the migration of the loading dye along the gel toward the red (positive) electrode.

28. Disconnect the battery pack when the loading dye band has run halfway off the end of the gel. *Note: If gels have been run overnight, this step will take place on the following day.*

29. Remove the cover from the electrophoresis chamber.

30. Carefully lift the gel tray (containing the gel) from the chamber onto a piece of paper towel. Notch one side of the gel so that you can identify the lanes. *Note: 0.8% agarose gels are very fragile, so use extreme care when handling the gel.*

31. Use a metric ruler to measure the distance of the dye bands in Lane 6 (in mm) from each of the six sample wells. *Note: Be sure to measure from the center of the well to the center of the band.*

Name _____ Class _____ Date _____

DNA Fingerprinting continued

32. Place the gel in a resealable band and add 1–2 mL of 1× TBE buffer and refrigerate the gel until the next lab period. If there is enough time (at least 45 minutes) left in the class period, proceed to step 33.

PART 3: STAINING THE GEL

After the DNA fragments have been separated, the gel can be stained to visualize the banding pattern of each DNA fingerprint.

33. Gently place the gel in the staining tray.

34. Wearing protective gloves, pour approximately 100 mL of warm dilute stain into the staining tray so that it covers the gel.

35. Cover the tray and let the gel stain for approximately 30 minutes.

36. Carefully decant the used stain. *Note: Make sure the gel remains flat and does not move up against the corner. Decant the stain directly to a sink drain and flush with water.*

37. Add distilled water or tap water to the staining tray. Do not pour water directly onto the gel.

38. Allow the gel to destain either overnight or for 20 minutes. To destain in 20 minutes, gently rock the tray, and change the water several times. Overnight destain does not require a change of water.

39. View the gel against a white sheet of paper. Sketch the bands you see on the blank gel in **Figure 2**.

40. Store the gel in a resealable plastic bag with 1–2 mL of 1× TBE buffer.

41. Dispose of your materials according to the directions from your teacher.

42. Clean up your work area and wash your hands before leaving the lab.

FIGURE 2 DNA FINGERPRINTS

Lane 1 DNA marker | Lane 2 Crime scene DNA | Lane 3 Suspect #1 DNA | Lane 4 Suspect #2 DNA | Lane 5 Suspect #3 DNA | Lane 6 Suspect #4 DNA

Name _____ Class _____ Date _____

DNA Fingerprinting *continued*

Analysis

1. **Explaining Events** Where are the smallest DNA fragments located on the gel? the largest? Explain why.

   _____
   _____
   _____
   _____
   _____
   _____

2. **Explaining Events** Why was loading dye added to each sample prior to loading the samples on the gel.

   _____
   _____
   _____
   _____
   _____
   _____

3. **Examining Data** Are any of the DNA fingerprints identical? If so, which ones?

   _____
   _____

Conclusions

1. **Drawing Conclusions** Based on your results, which suspect appears to be the burglar? Explain your answer.

   _____
   _____
   _____

Name _____ Class _____ Date _____

DNA Fingerprinting continued

2. Interpreting Information Explain why each suspect has a different banding pattern (or DNA fingerprint).

_____
_____
_____
_____
_____
_____

3. Making Predictions What do you think would happen if you placed your gel in the electrophoresis chamber with the wells containing DNA next to the red electrode instead of the black electrode?

_____
_____

4. Making Inferences From what types of crime scene evidence (other than blood) might police obtain DNA?

_____
_____

Extensions

1. **Research and Communications** Research another way that DNA fingerprinting is used in society and present your results to your classmates in a short oral report.
2. **Research and Communications** Research the procedures involved in collecting evidence at a crime scene.

Name _____ Class _____ Date _____

Skills Practice Lab

BIOTECHNOLOGY

Transforming Bacteria with a Firefly Gene

n 1928, scientists observed that certain bacteria could take up DNA, which had been released into the solution by other dead bacteria, and express the genes that were coded for by the DNA. They called this exchange of genetic material *transformation*. Cells that are able to take up DNA from their environment are termed *competent*. In many bacteria, certain environmental conditions can signal changes in gene expression that cause the cells to become competent. Many other cells, such as *E. coli*, do not become competent under ordinary conditions, but can be made competent by exposing them to a variety of artificial treatments, such as calcium ions.

The North American firefly contains a gene that codes for the production of the enzyme luciferase. Luciferase catalyzes a reaction that results in the luminescent glow of the firefly. The plasmid pBestLuc contains the gene that codes for luciferase as well as a gene that makes bacteria resistant to the antibiotic ampicillin. The plasmid pUC8 contains the gene for ampicillin resistance, but does not contain the luciferase gene.

In this lab, you will transform *E. coli* bacteria with the pBestLuc plasmid and determine if the transformation was successful by examining the cells for luminescence.

OBJECTIVES

Create competent cells by chemically and thermally treating *E. coli* cells.

Insert a plasmid containing firefly genes into competent E. coli cells.

Determine if transformation was successful by examining the cells for luminescence.

MATERIALS

- calcium chloride
- capillary micropipet (8) (or 200 µL micropipet)
- Bacti-spreader
- *E. coli* culture (in Luria agar tube)
- gloves
- graduated pipets, sterile (3)
- ice bath
- incubator
- inoculating loop, sterile (3)
- lab apron
- Luciferin solution (1 mM)
- Luria agar plate w/ampicillin (3)
- Luria broth
- microcentrifuge tube
- nitrocellulose membrane (2)
- pipet bulb, rubber
- plasmid pBestLuc®
- plasmid pUC8
- water bath (42° C)
- safety goggles

Name _____ Class _____ Date _____

Transforming Bacteria with a Firefly Gene *continued*

Procedure

1. Put on safety goggles, gloves, and a lab apron.
2. Obtain a tube of Luria agar with actively growing *E. coli*.
3. Obtain three microcentrifuge tubes. Label one tube "Luc," another tube "pUC" and the third tube "C" for control.
4. Using a 1 mL sterile graduated pipet, add 0.25 mL (250 μL) of ice-cold calcium chloride to each tube.
5. Using a sterile inoculating loop, transfer a large colony of the *E. coli* from the surface of the agar to each tube of calcium chloride. Gently move the loop across the surface of the agar to collect some of the bacteria. Be careful not to remove any agar with the loop.
6. Place the loop into the calcium chloride and twirl rapidly to remove the bacteria from the transfer loop. *Note: Gently tapping the loop against the side of the microcentrifuge tube may also help dislodge the bacteria.*
7. Dispose of the loop according to your teacher's instructions.
8. Add 10 μL of the plasmid pBestLuc solution to the tube labeled "Luc" by using the capillary micropipet and plunger.
9. Gently tap the tube with your finger to mix the plasmid into the solution. Place the tube on ice.
10. Use a different capillary micropipet to add 10 μL of the plasmid pUC8 solution to the tube labeled "pUC." Place the tube on ice.
11. Incubate all three tubes on ice for 15 minutes.
12. Obtain a Luria agar plate containing ampicillin while the tubes are incubating on ice.
13. Using forceps, or wearing latex gloves, place a piece of nitrocellulose membrane on the surface of the agar of two agar plates. Lift the lid of each plate only enough to place the membrane on the agar and replace the lid as soon as you are done. **CAUTION: Be very careful when handling the nitrocellulose membrane, it is very fragile.**
14. After 15 minutes, remove the tubes from the ice and heat shock them by immediately placing them in a 42° C waterbath for 75–90 seconds (in a foam tube holder). This procedure will allow the plasmids to enter the bacterial cells.
15. Remove the tubes from the water bath and immediately place on ice for 2 minutes.
16. Remove the tubes from the ice bath and add 0.25 mL (250 μL) of room temperature Luria broth to each tube using a different 1 mL sterile pipete for each tube.

Name _____ Class _____ Date _____

Transforming Bacteria with a Firefly Gene *continued*

17. Gently tap each tube with your finger to mix the solution. The tubes may now be kept at room temperature.

18. Label one of the agar plates containing nitrocellulose "Luc." Label the other agar plate containing nitrocellulose "pUC." Label the third agar plate "Control." Be sure to label all three plates with your group name.

19. Use a sterile pipet to add 0.20 mL (200 μL) of the solution in the "Luc" tube to the center of the agar plate with the nitrocellulose membrane labeled "Luc."

20. Gently rock the plate to ensure the solution thoroughly covers the surface of the membrane.

21. Using a new sterile pipet, add 0.20 mL (200 μL) of the solution in the "pUC" tube to the center of the agar plate with the nitrocellulose membrane labeled "pUC."

22. Gently rock the plate to ensure the solution thoroughly covers the surface of the membrane.

23. Using a new sterile pipet, add 0.20 mL (200 μL) of the solution in the "C" tube to the center of the agar plate labeled "Control."

24. Spread the cells over the entire surface of the agar using a sterile Bacti-spreader.

25. Allow the plates to sit undisturbed for one hour. After one hour, place the plates in a 37° C incubator, inverted, overnight.

26. After 24 hours, remove your plates from the incubator.

27. Count the approximate number of colonies growing on each plate and record the value below.

   Luc _____

   pUC _____

   Control _____

28. Remove the lid of the "Luc" and "pUC" plates and place them upside down on a flat surface. Make sure the lids are also labeled "Luc" and "pUC."

29. Add 0.5 mL of 1 mM (millimolar) luciferin to the center of each lid.

30. Carefully remove the nitrocellulose membrane containing the transformed colonies of *E. Coli* from each of the two agar plates and place in the appropriate lid on top of the luciferin solution.

31. Observe the luminescence in a dark area.

Name _____ Class _____ Date _____

Transforming Bacteria with a Firefly Gene *continued*

Analysis

1. Summarizing Data After taking your agar plates out of the incubator, what did you observe on each plate?

Luc _____

_____

_____

pUC _____

_____

_____

Control _____

_____

_____

2. Describing Events After placing the nitrocellulose membranes in the luciferin solution and visualizing them in a dark room, what did you observe?

Luc _____

_____

pUC _____

_____

_____

3. Explaining Events The control plate should not contain any bacterial colonies. Explain why.

_____

_____

_____

Name _____ Class _____ Date _____

Transforming Bacteria with a Firefly Gene *continued*

Conclusions

1. **Drawing Conclusions** Based on your observations, did transformation occur in each case? Explain.

   _____
   _____
   _____
   _____
   _____
   _____
   _____

2. **Interpreting Information** What role did placing the bacteria in a solution containing calcium chloride play in the transformation?

   _____
   _____

3. **Interpreting Information** Why was ampicillin added to the agar plates?

   _____
   _____

4. **Evaluating Results** Suppose bacterial colonies were present on the nitrocellulose membrane of the "Luc" plate, but no luminescence was observed. Form a hypothesis about what might cause this scenario.

   _____
   _____
   _____
   _____
   _____
   _____

Name _____ Class _____ Date _____

Transforming Bacteria with a Firefly Gene *continued*

Extensions

1. **Designing Experiments** Design an experiment that could be used to verify that transformation occurred in your experiment.

2. **Research and Communications** Research another type of genetic exchange that occurs in bacteria and briefly describe or draw the mechanism by which exchange occurs.

TEACHER RESOURCE PAGE

Name _____ Class _____ Date _____

| Quick Lab | DATASHEET FOR IN-TEXT LAB |

Modeling Gel Electrophoresis

You can use beads to model how DNA fragments are separated in a gel during electrophoresis.

MATERIALS
- 500 mL beaker
- large jar
- 3 sets of beads—each set a different size and different color

Procedure
1. Fill a large jar with the largest beads. The filled jar represents a gel.
2. Mix the two smaller beads in the beaker and then pour them slowly on top of the "gel." The two smaller size beads represent DNA fragments of different sizes.
3. Observe the flow of the beads through the "gel." Lightly agitate the jar if the beads do not flow easily.

Analysis
1. **Identify** which beads flowed through the "gel" the fastest.

   the smallest beads

2. **Relate** the sizes of the beads to the sizes of DNA fragments.

   The smaller beads represent the smaller DNA fragments.

3. **Determine** whether the top or the bottom of the jar represents the side of the gel with the positively charged pole.

   bottom; DNA is negatively charged and will flow to the pole with the

   opposite charge

4. **Critical Thinking**
   Forming Conclusions Why do the beads you identified in Analysis question 1 pass through the "gel" more quickly?

   The smaller size allows them to flow through the space in between the large

   beads.

Copyright © by Holt, Rinehart and Winston. All rights reserved.
Holt Biology   Gene Technology

TEACHER RESOURCE PAGE

Name _____ Class _____ Date _____

Exploration Lab

DATASHEET FOR IN-TEXT LAB

Modeling Recombinant DNA

SKILLS
- Modeling
- Comparing

OBJECTIVES
- **Construct** a model that can be used to explore the process of genetic engineering.
- **Describe** how recombinant DNA is made.

MATERIALS
- paper clips (56)
- plastic soda straw pieces (56)
- pushpins (15 red, 15 green, 13 blue, and 13 yellow)

Before You Begin

Genetic engineering is the process of taking a gene from one organism and inserting it into the DNA of another organism. The gene is delivered by a **vector**, such as a virus, or a bacterial **plasmid**.

First, a fragment of a chromosome that contains the gene is isolated by using a **restriction enzyme**, which cuts DNA at a specific nucleotide-base sequence. Some restriction enzymes cut DNA unevenly, producing single-stranded **sticky ends**. The DNA of the vector is cut by the same restriction enzyme. Next, the chromosome fragment is mixed with the cut DNA of the vector. Finally, an enzyme called **DNA ligase** joins the ends of the two types of cut DNA, producing **recombinant DNA.**

In this lab, you will model genetic engineering techniques. You will simulate the making of recombinant DNA that has a human gene inserted into the DNA of a plasmid.

1. Write a definition for each boldface term in the paragraph above and for the term *base-pairing rules*. Use a separate sheet of paper.
   Answers appear in the TE for this lab.

2. Based on the objectives for this lab, write a question you would like to explore about the process of genetic engineering.

   Answers will vary. For example: What steps are involved in transferring a

   gene from one organism to another one?

Copyright © by Holt, Rinehart and Winston. All rights reserved.

Holt Biology — Gene Technology

Modeling Recombinant DNA *continued*

Procedure
PART A: MODEL GENETIC ENGINEERING

1. Make 56 model nucleotides. To make a nucleotide, insert a pushpin midway along the length of a 3 cm piece of a soda straw. **CAUTION: Handle pushpins carefully. Pointed objects can cause injury.** Push a paper clip into one end of the soda-straw piece until it touches the pushpin.

2. Begin a model of a bacterial plasmid by arranging nucleotides for one DNA strand in the following order: blue, red, green, yellow, red, red, blue, blue, green, red, blue, green, red, blue, blue, green, yellow, and red. Join two adjacent nucleotides by inserting the paper clip end of one into the open end of the other.

3. Using your first DNA strand and the base-pairing rules, build the complementary strand of plasmid DNA. **Note:** *Yellow is complementary to blue, and green is complementary to red.*

4. Complete your model of a circular plasmid by joining the opposite ends of each DNA strand. Make a sketch showing the sequence of bases in your model plasmid. Use the abbreviations B, Y, G, and R for the pushpin colors. Your sketch should be similar to the one shown below.

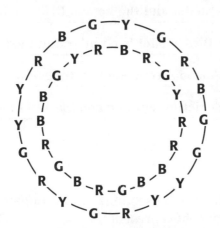

5. Begin a model of a human chromosome fragment made by a restriction enzyme. Place nucleotides for one DNA strand in the following order: BBRRYGGBRY. Build the second DNA strand by arranging the remaining nucleotides in the following order: BRRYGBYYGG.

6. Match the complementary portions of the two strands of DNA you made in step 5. Pair as many base pairs in a row as you can. Make a sketch showing the sequence of bases in your model of a human chromosome fragment.

7. Imagine that the restriction enzyme that cut the human chromosome fragment you made in steps 5 and 6 is moving around your model plasmid until it finds the sequence YRRBBG and its complementary sequence, BGGYYR. This restriction enzyme cuts each sequence between a B and a G. Find such a section in your sketch of your model plasmid's DNA.

Name _____ Class _____ Date _____

Modeling Recombinant DNA continued

8. Simulate the action of the restriction enzyme on the section you identified in step 7. Open both strands of your model plasmid's DNA by pulling apart the adjacent green and blue nucleotides in each strand. Make a sketch of the split plasmid DNA molecule.

9. Move your model human DNA fragment into the break in your model plasmid's DNA molecule. Imagine that a ligase joins the ends of the human and plasmid DNA. Make a sketch of your final model DNA molecule.

PART B: CLEANUP AND DISPOSAL

10. Dispose of damaged pushpins in the designated waste container.

11. Clean up your work area and all lab equipment. Return lab equipment to its proper place. Wash your hands thoroughly before you leave the lab and after you finish all work.

Analyze and Conclude

1. **Comparing Structures** Compare your models of plasmid DNA and human DNA.

    Both the plasmid DNA model and the human DNA model are double-stranded. The plasmid DNA model is circular and the human DNA model is linear. The human DNA model represents a fragment of a DNA molecule that has been cut with a restriction enzyme and has sticky ends.

2. **Relating Concepts** What do the sections of four unpaired nucleotides in your model human DNA fragment represent?

    sticky ends

3. **Comparing Structures** How did your original model plasmid DNA molecule differ from your final model DNA molecule?

    The original model plasmid DNA was smaller (had fewer nucleotides) and did not contain a gene from a human chromosome.

Modeling Recombinant DNA *continued*

4. Drawing Conclusions What does the molecule you made in step 9 represent?

recombinant DNA

5. Further Inquiry Write a new question that could be explored with another investigation.

Answers will vary. For example: What happens if the plasmid and human DNA do not have complementary sticky ends?

TEACHER RESOURCE PAGE

Exploration Lab

DNA Fingerprinting

BIOTECHNOLOGY

Teacher Notes

TIME REQUIRED Three 45-minute periods and 15 minutes on another day.

SKILLS ACQUIRED

Collecting data
Experimenting
Identifying patterns
Recognizing patterns
Inferring
Interpreting
Analyzing data
Predicting

RATINGS Easy ← 1 2 3 4 → Hard

Teacher Prep–2
Student Setup–2
Concept Level–4
Cleanup–2

THE SCIENTIFIC METHOD

Make Observations Students observe simulated DNA fingerprints.

Form a Hypothesis Students form a hypothesis in Conclusions question 3.

Analyze the Results Analysis questions 1–3 require students to analyze their results.

Draw Conclusions Conclusions question 1 asks student to draw conclusions from their data.

MATERIALS

Materials for this lab can be purchased as a kit from WARD'S. See the *Master Materials List* for ordering instructions.

Precast gels come with the WARD'S kit. To cast 0.8% agarose gels, refer to product literature supplied with prepared agarose. Prepare 1× TBE running buffer according to manufacturer's instructions.

A 37° C incubator may be used in place of the 37° C water bath.

SAFETY CAUTIONS

- Discuss all safety symbols and caution statements with students.

- Be sure that students know the correct procedure for making electrical connections. Students should be supervised at all times during this investigation.

- Do not come in personal contact with or allow metal or any conductive material to come in contact with the reservoir buffer or the electrophorectic cell while the battery/power supply is on.

- If using a power supply, be sure to follow manufacturer's directions and precautions.

Copyright © by Holt, Rinehart and Winston. All rights reserved.

Holt Biology   Gene Technology

DNA Fingerprinting continued

DISPOSAL

DNA stain, DNA samples and 1× TBE buffer can be flushed down the drain with copious amounts of water. Stained agarose gels and other materials used in this lab can be put in the trash.

TECHNIQUES TO DEMONSTRATE

Demonstrate how to use wrist action to snap the tube so that the liquid is forced to the bottom of the tube. Also, demonstrate how to use a micropipet to load samples into gel wells. Caution students to be very careful not to puncture the gel with the micropipet tips when loading samples.

TIPS AND TRICKS

This lab works best in groups of three to six students.

You may pour agarose gels in advance and store them in the refrigerator for up to one week prior to the electrophoresis portion of the lab. Place the gel, on a casting tray, in a resealable bag with 1–2 mL 1× TBE buffer.

Remove DNA samples and enzymes from the freezer and allow them to thaw 15 minutes prior to performing the lab. After samples have thawed, tap them on the countertop to make sure all of the liquid is on the bottom of the tube and not in the cap of the vial. Place and keep samples on ice.

To visualize some of the DNA as the gel is being run, prestain the gel and buffer by adding 1 μL of DNA stain concentrate per mL of 1× TBE running buffer and by adding 1 μL of DNA stain concentrate per mL of liquefied agarose. Staining and destaining following electrophoresis will still be necessary.

Gels may also be loaded *after* the 1× TBE running buffer has been added. Loading the gel "wet" decreases the likelihood that the dye samples will be washed out of the wells, but makes it harder to see the wells and to avoid puncturing the bottom of the wells when loading. To "wet" load the gel, place a strip of black or dark colored paper under the electrophoresis chamber so that it is under the wells. Add 1× TBE running buffer until it covers the gel by 2–3 mm. Use a micropipet to load each well. Do not overload the wells.

1× TBE buffer may be stored at room temperature for up to 18 months. If the concentrated buffer contains a white precipitate, it should be discarded.

When using a battery operated electrophoresis chamber, it will take up to two hours for the DNA to separate. If you use three batteries instead of five, the DNA can be separated overnight. For shorter running times (under an hour), use a power supply set between 75 V and 125 V.

Prepare stain immediately before staining by adding stain concentrate to warm (50–55° C) water. Warm stain results in enhanced banding of the DNA fragments. Dilute stain may be saved and reused several times. For best results, reheat the stain before reusing.

For long-term storage of gels (up to a month), add several drops of dilute stain to the 1× TBE in the bag to prevent DNA bands from fading. Gels can be restained if necessary.

Exploration Lab
DNA Fingerprinting

BIOTECHNOLOGY

With the exception of identical twins, no two people have the same DNA sequence. The slight differences (or polymorphisms) in each person's DNA sequence ensure that each person (except identical twins) has a unique set of sites where restriction enzymes can bind and cut the DNA. These differences are know as *restriction fragment length polymorphisms (RFLPs)*. Due to RFLPs, the exact number and size of fragments produced by digestion with a specific restriction enzyme varies from person to person. When the fragments are separated on a gel by electrophoresis, a unique banding pattern results that is known as a DNA fingerprint. An unknown DNA sample can be matched with a known sample of a person's DNA to an accuracy of 1 in 10 billion people. Due to its accuracy, DNA fingerprinting is often used in the legal system to determine identity in criminal cases and to establish hereditary relationships, such as paternity.

In this lab, you will perform some of the experimental procedures involved in DNA fingerprinting and use your results to identify a hypothetical burglar.

OBJECTIVES

Conduct a restriction digestion of DNA samples.

Use agarose gel electrophoresis to separate the digested DNA samples.

Evaluate the results of simulated DNA fingerprints.

Identify a hypothetical burglar by analyzing simulated DNA fingerprints.

MATERIALS

- agarose gel on gel tray (0.8%)
- beaker, 500 mL
- DNA samples (5)
- DNA stain
- electrophoresis system, battery powered
- gel staining tray
- gloves
- graduated cylinder, 250 mL
- lab apron
- Lambda DNA/*Hind III* marker
- loading dye
- microfuge tubes (6)
- micropipet, 10 or 20 µL
- micropipet tips (30)
- *Pvu II* restriction enzyme
- *Pvu II* reaction buffer, 10x
- microfuge tube rack/float
- TBE running buffer, 350 mL (1×)
- safety goggles
- water bath, 37° C
- water bath, 65° C

TEACHER RESOURCE PAGE

Name _____ Class _____ Date _____

DNA Fingerprinting *continued*

Procedure

1. Read the following scenario.

   The police are investigating a burglary. The police report indicates that as the police arrived, the masked burglar smashed a glass door with a chair and quickly escaped. Police found small pieces of bloodstained fabric on some of the pieces of the glass door. Four brothers have been brought in as suspects. Because all of them have the same blood type, investigators have resorted to DNA fingerprinting to identify which brother is the burglar. You have been provided with the DNA from the blood samples found at the crime scene and DNA from blood samples taken from each of the four suspects.

PART 1: RESTRICTION ENZYME DIGESTION OF DNA

In the first step of DNA fingerprinting, known and unknown samples are obtained and then digested, or cut into small fragments (RFLPs), by the same restriction enzyme.

2. Put on safety goggles and a lab apron.

3. Obtain five microcentrifuge tubes and label as follows: Crime Scene, Suspect #1, Suspect #2, Suspect #3, Suspect #4.

4. Set the micropipet to 2 μL and place a clean tip on the pipet.

5. Add 2 μL of *Pvu* II 10x reaction buffer to each tube.

6. Set the micropipet to 10 μL.

7. Using a fresh pipet tip for each sample, add 10 μL of each DNA sample to the appropriate tube.

8. Set the micropipet to 2 μL.

9. Add 2 μL of Pvu II restriction enzyme to each tube. Again, be sure to use a fresh micropipet tip for each sample. *Note: It may be helpful to place the pipet tip against the side of the microcentrifuge tube when dispensing small volumes.*

10. Close each tube and snap the tubes in a downward motion with your wrist to force all of the reagents to the bottom of the tubes.

11. Incubate all five tubes at 37° C for 45–60 minutes.

12. Add 2 μL of loading dye to each tube.

13. Incubate all five tubes for five minutes at 65° C to stop the restriction enzyme activity.

14. Place the samples in the freezer if you will not be using them immediately. Make sure all samples for your group are stored together and labeled with your group name.

Copyright © by Holt, Rinehart and Winston. All rights reserved.

Holt Biology — Gene Technology

Name _____ Class _____ Date _____

DNA Fingerprinting continued

PART 2: GEL ELECTROPHORESIS OF DNA SAMPLES

The next step of a DNA fingerprint is to separate the RFLPs by gel electrophoresis. In gel electrophoresis, the digested DNA samples are loaded into wells on a jellylike slab called a gel. The gel is then exposed to an electrical current. DNA has a negative electrical charge, so the RFLPs are attracted to the positive pole when an electric current is applied. Smaller fragments travel farther through the gel than longer ones. A special dye, called loading dye, is added to the DNA samples for two reasons. The loading dye is heavier than the buffer, which helps the samples to stay in the wells. The loading dye also runs slightly faster than the DNA samples and indicates when the gel has finished running.

15. Place the tray with the gel on the lab bench.

16. Set the pipet to 10 μL and place a new tip on the end of your micropipet.

17. Open the microtube containing the DNA marker and use the pipet to load 10 μL of the marker DNA into the well in Lane 1 of an agarose gel. To do this, place both elbows on the lab table, lean over the gel, and slowly lower the micropipet tip into the opening of the well before depressing the plunger. *Note: Do not jab the micropipet tip through the bottom of the well.*

18. Set the pipet to 15 μL and place a new tip on the end of your micropipet.

19. Open the microtube containing the crime scene sample and use the pipet to load 15 μL of the sample into the well in Lane 2 of an agarose gel.

20. Using a new micropipet tip for each tube, repeat step 19 for each of the remaining samples. Place each sample in a well, according to the following lane assignments:

    Lane 1   DNA marker
    Lane 2   Crime scene DNA
    Lane 3   Suspect #1 DNA
    Lane 4   Suspect #2 DNA
    Lane 5   Suspect #3 DNA
    Lane 6   Suspect #4 DNA

21. Carefully place the agarose gel (still in a gel-casting tray) in the electrophoresis chamber of an electrophoresis apparatus, shown in **Figure 1,** so that the wells are closest to the negative electrode.

22. Pour approximately 350 mL of 1×TBE running buffer into a beaker.

23. *Gently* and *slowly* pour the running buffer from the beaker into one side of the electrophoresis chamber until the gel is completely covered (approximately 1 to 2 mm above the top surface of the gel). *Note: Pouring too fast will rinse your DNA sample out of the wells. Be careful not to overfill the chamber with buffer.*

DNA Fingerprinting continued

FIGURE 1 SETTING UP THE ELECTROPHORESIS APPARATUS

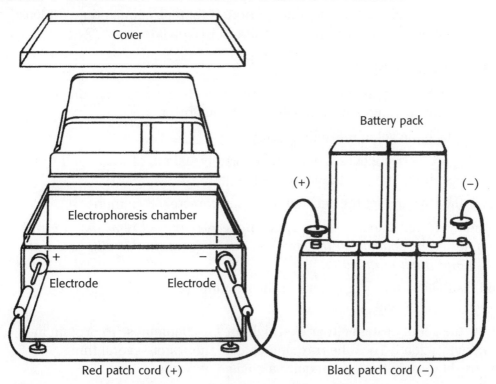

24. Place the cover on the electrophoresis chamber. Wipe off any spills around the electrophoretic apparatus before doing the next step.

25. Connect five 9 V alkaline batteries as shown in **Figure 1. CAUTION: Do not touch both ends of the patch cords or both terminals on the battery pack at the same time.**

26. Connect the red (positive) patch cord to the red terminal on the chamber and the red terminal on the battery pack. Follow the same procedure with the black (negative) patch cord and the black terminals.

27. Observe the migration of the loading dye along the gel toward the red (positive) electrode.

28. Disconnect the battery pack when the loading dye band has run halfway off the end of the gel. *Note: If gels have been run overnight, this step will take place on the following day.*

29. Remove the cover from the electrophoresis chamber.

30. Carefully lift the gel tray (containing the gel) from the chamber onto a piece of paper towel. Notch one side of the gel so that you can identify the lanes. *Note: 0.8% agarose gels are very fragile, so use extreme care when handling the gel.*

31. Use a metric ruler to measure the distance of the dye bands in Lane 6 (in mm) from each of the six sample wells. *Note: Be sure to measure from the center of the well to the center of the band.*

Name _____ Class _____ Date _____

DNA Fingerprinting continued

32. Place the gel in a resealable band and add 1–2 mL of 1× TBE buffer and refrigerate the gel until the next lab period. If there is enough time (at least 45 minutes) left in the class period, proceed to step 33.

PART 3: STAINING THE GEL

After the DNA fragments have been separated, the gel can be stained to visualize the banding pattern of each DNA fingerprint.

33. Gently place the gel in the staining tray.
34. Wearing protective gloves, pour approximately 100 mL of warm dilute stain into the staining tray so that it covers the gel.
35. Cover the tray and let the gel stain for approximately 30 minutes.
36. Carefully decant the used stain. *Note: Make sure the gel remains flat and does not move up against the corner. Decant the stain directly to a sink drain and flush with water.*
37. Add distilled water or tap water to the staining tray. Do not pour water directly onto the gel.
38. Allow the gel to destain either overnight or for 20 minutes. To destain in 20 minutes, gently rock the tray, and change the water several times. Overnight destain does not require a change of water.
39. View the gel against a white sheet of paper. Sketch the bands you see on the blank gel in **Figure 2**.
40. Store the gel in a resealable plastic bag with 1–2 mL of 1× TBE buffer.
41. Dispose of your materials according to the directions from your teacher.
42. Clean up your work area and wash your hands before leaving the lab.

FIGURE 2 DNA FINGERPRINTS

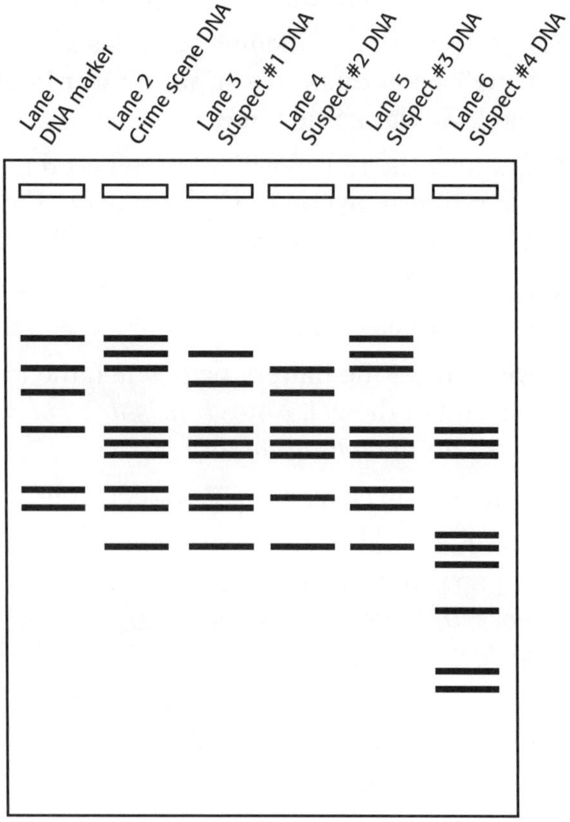

Lane 1 DNA marker | Lane 2 Crime scene DNA | Lane 3 Suspect #1 DNA | Lane 4 Suspect #2 DNA | Lane 5 Suspect #3 DNA | Lane 6 Suspect #4 DNA

TEACHER RESOURCE PAGE

Name _____ Class _____ Date _____

DNA Fingerprinting continued

Analysis

1. Explaining Events Where are the smallest DNA fragments located on the gel? the largest? Explain why.

The smallest DNA fragments are the bands located near the bottom of the gel. The largest DNA fragments are the bands located near the wells. All of the DNA has a negative charge and is attracted to the positive red anode. Because the smaller fragments move more quickly and easily through the agarose than the larger fragments, they are able to move farther down the gel.

2. Explaining Events Why was loading dye added to each sample prior to loading the samples on the gel.

Answers may vary, but students should suggest that the loading dye runs slightly faster than the DNA samples and indicates when the DNA fragment have separated and the gel is done running. DNA loading dye is also heavier than the electrophoresis buffer, which helps the DNA samples to stay in the wells.

3. Examining Data Are any of the DNA fingerprints identical? If so, which ones?

The DNA fingerprint from the crime scene DNA should match the DNA fingerprint of Suspect #3.

Conclusions

1. Drawing Conclusions Based on your results, which suspect appears to be the burglar? Explain your answer.

Suspect # 3, because the DNA fingerprint from the crime scene DNA matches the DNA fingerprint of Suspect #3.

Copyright © by Holt, Rinehart and Winston. All rights reserved.
Holt Biology   Gene Technology

TEACHER RESOURCE PAGE

Name _____ Class _____ Date _____

DNA Fingerprinting continued

2. **Interpreting Information** Explain why each suspect has a different banding pattern (or DNA fingerprint).

   The slight differences (or RFLPs) in each person's DNA sequence ensure that each person (except identical twins) has a unique set of sites where restriction enzymes can bind and cut the DNA. Due to RFLPs, the exact number and size of fragments produced by digestion with a specific restriction enzyme varies from person to person. Separation of the RFLPs by gel electrophoresis results in a unique banding pattern.

3. **Making Predictions** What do you think would happen if you placed your gel in the electrophoresis chamber with the wells containing DNA next to the red electrode instead of the black electrode?

   The DNA would be attracted to the red anode and would quickly run off the top of the gel and into the buffer.

4. **Making Inferences** From what types of crime scene evidence (other than blood) might police obtain DNA?

   Answers will vary. DNA could be extracted from skin, hair, semen, and saliva.

Extensions

1. **Research and Communications** Research another way that DNA fingerprinting is used in society and present your results to your classmates in a short oral report.
2. **Research and Communications** Research the procedures involved in collecting evidence at a crime scene.

TEACHER RESOURCE PAGE

Skills Practice Lab

BIOTECHNOLOGY

Transforming Bacteria with a Firefly Gene

Teacher Notes

TIME REQUIRED One 45-minute period, then 20 minutes on a second day.

SKILLS ACQUIRED
Experimenting
Identifying patterns
Interpreting
Analyzing data
Predicting

RATINGS

Easy ← 1 2 3 4 → Hard

Teacher Prep–4
Student Setup–3
Concept Level–3
Cleanup–2

THE SCIENTIFIC METHOD

Make Observations Students observe cells for luminescence.

Form a Hypothesis Students form a hypothesis in Conclusions question 4.

Analyze the Results Analysis question 3 and Conclusions questions 1 and 4 require students to analyze their results.

Draw Conclusions Students draw conclusions in Conclusions question 1.

MATERIALS

Materials for this lab can be purchased as a kit from WARD'S. See the *Master Materials List* for ordering instructions.

The WARD's Glowing Bacteria kit contains many of the materials necessary for this lab. To prepare the solutions, media, and cultures for this lab, follow the instructions enclosed with the kit. Additional materials needed include the following: autoclave, bunsen burner and striker, hot plate, insulated gloves, saucepan or beaker, thermometer.

SAFETY CAUTIONS

- Discuss all safety symbols and caution statements with students.

- Consult with the school nurse to screen students whose immune system may be compromised by illness or who may be receiving immunosupressive drug therapy. Such individuals are extraordinarily sensitive to potential infection from generally harmless microorgansims and should not participate in this lab. Do not allow students with any open cuts, abrasions, or sores to work with microorganisms.

Copyright © by Holt, Rinehart and Winston. All rights reserved.

Holt Biology　　　　　　　　　　　　　　　Gene Technology

Transforming Bacteria with a Firefly Gene continued

- Treat ALL microbes as pathenogenic. Seal with tape all petri dishes containing bacterial cultures.
- Never allow students to clean up bacteriological spills. Keep on hand a spill kit containing 500 mL of full-strength household bleach, biohazard bags (autoclavable), forceps, and paper towels.
- In the event of a bacterial spill, cover the area with a layer of paper towels. Wet the paper towels with the bleach, and allow to stand for 15–20 minutes. Wearing gloves and using forceps, place the residue in the biohazard bag. If broken glass is present, use a brush and dustpan to collect the broken material and place it in a suitably marked container.
- Wash all lab surfaces with a disinfectant solution before and after handling bacterial cultures.

DISPOSAL

Autoclave or steam-sterilize all used cultures and any materials that have come into contact with them at 120° C and 15 psi for 15–20 minutes. If an autoclave or steam-sterilizer is not available, flood or immerse these articles with full strength household bleach for 30 minutes, and then discard. Place disposal bags, following autoclaving, in an outer-sealed container with a lid. Make sure all free liquids are absorbed with paper towels to minimize leakage. Contaminated materials that are to be decontaminated away from the laboratory must be placed in a durable, leakproof container that is closed prior to removal from the laboratory.

TECHNIQUES TO DEMONSTRATE

Demonstrate the correct use of aseptic technique prior to conducting the lab. Also demonstrate how to use the innoculating loop to carefully collect a colony of bacteria.

TIPS AND TRICKS

This lab works best in groups of three to five students.

Sterile technique should be used throughout the pre-lab preparation and procedure.

The calcium chloride and plasmid pBestLuc should be in an ice bath and kept cold throughout the experiment.

You may wish to have the students prepare the antibiotic mixture and pour their own agar plates the day before the lab, or you may prepare them yourself in advance.

Set up 42° C hot-water baths for students ahead of time.

Make sure to allow enough time for students eyes to adjust to the dark when observing luminescence.

TEACHER RESOURCE PAGE

Name _____ Class _____ Date _____

Skills Practice Lab   BIOTECHNOLOGY

Transforming Bacteria with a Firefly Gene

In 1928, scientists observed that certain bacteria could take up DNA, which had been released into the solution by other dead bacteria, and express the genes that were coded for by the DNA. They called this exchange of genetic material *transformation*. Cells that are able to take up DNA from their environment are termed *competent*. In many bacteria, certain environmental conditions can signal changes in gene expression that cause the cells to become competent. Many other cells, such as *E. coli*, do not become competent under ordinary conditions, but can be made competent by exposing them to a variety of artificial treatments, such as calcium ions.

The North American firefly contains a gene that codes for the production of the enzyme luciferase. Luciferase catalyzes a reaction that results in the luminescent glow of the firefly. The plasmid pBestLuc contains the gene that codes for luciferase as well as a gene that makes bacteria resistant to the antibiotic ampicillin. The plasmid pUC8 contains the gene for ampicillin resistance, but does not contain the luciferase gene.

In this lab, you will transform *E. coli* bacteria with the pBestLuc plasmid and determine if the transformation was successful by examining the cells for luminescence.

OBJECTIVES

Create competent cells by chemically and thermally treating *E. coli* cells.

Insert a plasmid containing firefly genes into competent E. coli cells.

Determine if transformation was successful by examining the cells for luminescence.

MATERIALS

- calcium chloride
- capillary micropipet (8) (or 200 µL micropipet)
- Bacti-spreader
- *E. coli* culture (in Luria agar tube)
- gloves
- graduated pipets, sterile (3)
- ice bath
- incubator
- inoculating loop, sterile (3)
- lab apron
- Luciferin solution (1 mM)
- Luria agar plate w/ampicillin (3)
- Luria broth
- microcentrifuge tube
- nitrocellulose membrane (2)
- pipet bulb, rubber
- plasmid pBestLuc®
- plasmid pUC8
- water bath (42° C)
- safety goggles

Transforming Bacteria with a Firefly Gene *continued*

Procedure

1. Put on safety goggles, gloves, and a lab apron.
2. Obtain a tube of Luria agar with actively growing *E. coli*.
3. Obtain three microcentrifuge tubes. Label one tube "Luc," another tube "pUC" and the third tube "C" for control.
4. Using a 1 mL sterile graduated pipet, add 0.25 mL (250 μL) of ice-cold calcium chloride to each tube.
5. Using a sterile inoculating loop, transfer a large colony of the *E. coli* from the surface of the agar to each tube of calcium chloride. Gently move the loop across the surface of the agar to collect some of the bacteria. Be careful not to remove any agar with the loop.
6. Place the loop into the calcium chloride and twirl rapidly to remove the bacteria from the transfer loop. *Note: Gently tapping the loop against the side of the microcentrifuge tube may also help dislodge the bacteria.*
7. Dispose of the loop according to your teacher's instructions.
8. Add 10 μL of the plasmid pBestLuc solution to the tube labeled "Luc" by using the capillary micropipet and plunger.
9. Gently tap the tube with your finger to mix the plasmid into the solution. Place the tube on ice.
10. Use a different capillary micropipet to add 10 μL of the plasmid pUC8 solution to the tube labeled "pUC." Place the tube on ice.
11. Incubate all three tubes on ice for 15 minutes.
12. Obtain a Luria agar plate containing ampicillin while the tubes are incubating on ice.
13. Using forceps, or wearing latex gloves, place a piece of nitrocellulose membrane on the surface of the agar of two agar plates. Lift the lid of each plate only enough to place the membrane on the agar and replace the lid as soon as you are done. **CAUTION: Be very careful when handling the nitrocellulose membrane, it is very fragile.**
14. After 15 minutes, remove the tubes from the ice and heat shock them by immediately placing them in a 42° C waterbath for 75–90 seconds (in a foam tube holder). This procedure will allow the plasmids to enter the bacterial cells.
15. Remove the tubes from the water bath and immediately place on ice for 2 minutes.
16. Remove the tubes from the ice bath and add 0.25 mL (250 μL) of room temperature Luria broth to each tube using a different 1 mL sterile pipete for each tube.

TEACHER RESOURCE PAGE

Name _____ Class _____ Date _____

| Transforming Bacteria with a Firefly Gene *continued*

17. Gently tap each tube with your finger to mix the solution. The tubes may now be kept at room temperature.

18. Label one of the agar plates containing nitrocellulose "Luc." Label the other agar plate containing nitrocellulose "pUC." Label the third agar plate "Control." Be sure to label all three plates with your group name.

19. Use a sterile pipet to add 0.20 mL (200 µL) of the solution in the "Luc" tube to the center of the agar plate with the nitrocellulose membrane labeled "Luc."

20. Gently rock the plate to ensure the solution thoroughly covers the surface of the membrane.

21. Using a new sterile pipet, add 0.20 mL (200 µL) of the solution in the "pUC" tube to the center of the agar plate with the nitrocellulose membrane labeled "pUC."

22. Gently rock the plate to ensure the solution thoroughly covers the surface of the membrane.

23. Using a new sterile pipet, add 0.20 mL (200 µL) of the solution in the "C" tube to the center of the agar plate labeled "Control."

24. Spread the cells over the entire surface of the agar using a sterile Bacti-spreader.

25. Allow the plates to sit undisturbed for one hour. After one hour, place the plates in a 37° C incubator, inverted, overnight.

26. After 24 hours, remove your plates from the incubator.

27. Count the approximate number of colonies growing on each plate and record the value below.

Luc _____

pUC _____

Control _____

28. Remove the lid of the "Luc" and "pUC" plates and place them upside down on a flat surface. Make sure the lids are also labeled "Luc" and "pUC."

29. Add 0.5 mL of 1 mM (millimolar) luciferin to the center of each lid.

30. Carefully remove the nitrocellulose membrane containing the transformed colonies of *E. Coli* from each of the two agar plates and place in the appropriate lid on top of the luciferin solution.

31. Observe the luminescence in a dark area.

TEACHER RESOURCE PAGE

Name _____ Class _____ Date _____

Transforming Bacteria with a Firefly Gene *continued*

Analysis

1. Summarizing Data After taking your agar plates out of the incubator, what did you observe on each plate?

Luc _Answers will vary. If the transformation was successful, students should observe between 20 and 300 bacterial colonies on the surface of the nitrocellulose._

pUC _Answers will vary. If the transformation was successful, students should observe between 20 and 300 bacterial colonies on the surface of the nitrocellulose._

Control _Answers will vary. Students should not observe any bacterial colonies on the agar. Presence of bacteria probably indicates that cross-contamination occurred._

2. Describing Events After placing the nitrocellulose membranes in the luciferin solution and visualizing them in a dark room, what did you observe?

Luc _If the transformation was successful, students should observe that the bacterial colonies exhibit luminescence. (They glow in the dark.)_

pUC _Bacterial colonies on this nitrocellulose membrane should NOT exhibit luminescence._

3. Explaining Events The control plate should not contain any bacterial colonies. Explain why.

All of the bacteria spread onto this plate should have been killed by the antibiotic ampicillin. The bacteria died because the gene that confers resistance to the antibiotic ampicillin was not transformed into the control bacteria.

Name _____ Class _____ Date _____

Transforming Bacteria with a Firefly Gene *continued*

Conclusions

1. **Drawing Conclusions** Based on your observations, did transformation occur in each case? Explain.

   Answers may vary. For "Luc," the presence of luminescent bacterial colonies indicates that transformation occurred. For "pUC," the presence of non-luminescent bacterial colonies indicates that transformation occurred. For "Control," the absence of bacterial colonies indicates that transformation did not occur. The presence of bacterial colonies on the control plate probably indicates cross-contamination from poor technique because plasmid was not added to this condition.

2. **Interpreting Information** What role did placing the bacteria in a solution containing calcium chloride play in the transformation?

   The calcium chloride helped the cells to become competent (able to take up the plasmid).

3. **Interpreting Information** Why was ampicillin added to the agar plates?

   The presence of ampicillin in the agar plates allowed the growth of only those bacteria that had taken up the plasmid containing the gene for ampicillin resistance.

4. **Evaluating Results** Suppose bacterial colonies were present on the nitrocellulose membrane of the "Luc" plate, but no luminescence was observed. Form a hypothesis about what might cause this scenario.

   Answers will vary. Accept all reasonable answers. 1) It's possible that the cells were transformed with pBestLuc, but that the luminescence was too weak to visualize. 2) Ampicillin might have been missing altogether or the amount might have been inadequate, allowing non-transformed bacteria to grow on the plates. 3) The bacteria might have been transformed with a plasmid containing the ampicillin resistance gene other than the pBestLuc plasmid (such as the pUC8 plasmid), and thus been able to grow on the ampicillin plates.

Copyright © by Holt, Rinehart and Winston. All rights reserved.
Holt Biology — Gene Technology

Name _____ Class _____ Date _____

Transforming Bacteria with a Firefly Gene continued

Extensions

1. **Designing Experiments** Design an experiment that could be used to verify that transformation occurred in your experiment.

2. **Research and Communications** Research another type of genetic exchange that occurs in bacteria and briefly describe or draw the mechanism by which exchange occurs.

Answer Key

Directed Reading

SECTION: GENETIC ENGINEERING
1. DNA
2. bacterial
3. genetic
4. Restriction
5. plasmids
6. ligase
7. cloning
8. sticky
9. 3
10. 1
11. 4
12. 2
13. restriction enzymes
14. electrophoresis
15. negatively
16. less
17. probe

SECTION: HUMAN APPLICATIONS OF GENETIC ENGINEERING
1. the Human Genome Project
2. 30,000
3. 3 billion
4. insulin, hemophilia
5. vaccine
6. proteins, antibodies
7. Viruses, such as HIV, can be transferred along with the proteins.
8. Vaccines made from killed or weakened viruses may transmit a disease inadvertently if the process by which they are made fails to kill or weaken all the viruses. The genetically engineered herpes II vaccine cannot cause the disease because the only part of the virus in the vaccine is the gene that codes for a surface protein.
9. Protein factor VIII is derived from human blood, which can be infected with HIV or the hepatitis B virus.
10. A DNA fingerprint is a pattern of dark bands made on photographic film.
11. A DNA fingerprint is made when an individual's DNA is cut into fragments (RFLPs) by restriction enzymes. The restriction fragments are then separated by electrophoresis, probed, and finally exposed to X-ray film.
12. Each individual has unique genetic material, which produces a unique DNA fingerprint.

SECTION: GENETIC ENGINEERING IN AGRICULTURE
1. genes
2. weedkiller
3. pesticides
4. weeds
5. Producing growth hormones with bacteria is inexpensive, making it practical to add growth hormones as a supplement to the cows' diets.
6. to produce human proteins that can be used for pharmaceutical purposes
7. Transgenic animals are animals that have foreign DNA in their cells.
8. First, he removed mammary cells from a sheep and grew them in a solution that stops the cell cycle. Next, he extracted egg cells from another sheep and removed the nucleus from each cell. Then he placed the mammary cells next to the empty egg cells and exposed the cells to an electric shock, which fused the mammary cells with the egg cells and triggered cell division in the fused cells. After a fused cell divided and formed an embryo, he implanted the embryo into a surrogate mother.

Active Reading

SECTION: GENETIC ENGINEERING
1. bacterial enzymes that recognize and bind to specific short sequences of DNA
2. an agent that is used to carry the gene of interest into another cell
3. Plasmids are circular DNA molecules that are commonly used as vectors.
4. DNA ligase is an enzyme that is added to DNA fragments to help them bond together.
5. Gene cloning produces many copies of the gene of interest.

TEACHER RESOURCE PAGE

6. When bacteria reproduce by binary fission, offspring that are identical to the solitary parent are produced. When a bacterial cell replicates its DNA, it also replicates its plasmid DNA.
7. c

SECTION: HUMAN APPLICATIONS OF GENETIC ENGINEERING
1. smallpox and polio
2. a solution containing a harmless version of a pathogen
3. The words *disease-causing microorganism* define the term *pathogen*, which precedes the parentheses.
4. This action is caused by injection of a vaccine into the body.
5. The immune system makes antibodies.
6. The vaccine is prepared either by killing a specific pathogenic microbe or by making the microbe unable to grow.
7. b

SECTION: GENETIC ENGINEERING IN AGRICULTURE
1. the first successful cloning using differentiated cells from an adult animal
2. a cell that has become specialized to become a specific type of cell
3. It clarifies the term *udder*, which precedes the parentheses.
4. that only embryonic or fetal cells could be cloned and that differentiated cells could not give rise to an entire organism
5. 3
6. 1
7. 5
8. 4
9. 2
10. a

Vocabulary Review

1. c; Human Genome Project
2. h; vector
3. g; plasmid
4. i; recombinant DNA
5. b; vaccine
6. e; restriction enzymes
7. d; transgenic animal
8. j; gene cloning
9. f; electrophoresis
10. a; transgenic animal
11. k; genetic engineering

Science Skills

INTERPRETING DIAGRAMS
1. The structure labeled A is a plasmid. It is removed from the bacterial cell so that the plasmid can be used as a vector to carry the insulin gene into a bacterial cell.
2. Restriction enzymes are used to cut DNA. DNA molecules cut with restriction enzymes have sticky ends that allow different DNA fragments cut with the same restriction enzyme to combine.
3. This is recombinant DNA. (The DNA with the insulin gene and plasmid DNA are combined).
4. This is a bacterial cell that contains recombined plasmids (plasmids containing the insulin gene).
5. CCGG and GGCC
6. pair on the left—GGCC/CCGG; pair on the right—CCGG/GGCC
7. Tetracycline, an antibiotic, destroys bacterial cells. Some bacterial cells, however, contain a gene for tetracycline resistance in their plasmid DNA, and they are not harmed by the antibiotic. These cells are called tetracycline-resistant cells. In the diagram, the plasmid DNA used in the genetic engineering experiment has the gene for tetracycline resistance. Only the cells that have taken up the plasmid DNA with the gene for tetracycline resistance survive when tetracycline is added to the cultures. Only the resistant cells, those that also carry the gene of interest, survive.

Concept Mapping

1. genetic engineering
2. agriculture
3. electrophoresis, probes, or restriction enzymes
4. probes, electrophoresis, or restriction enzymes
5. restriction enzymes, electrophoresis, or probes
6. vaccines or drugs
7. drugs or vaccines
8. cloned animals

Critical Thinking

1. b
2. a
3. d
4. c
5. c
6. e
7. a
8. d
9. b
10. b
11. d
12. a
13. c
14. d, b
15. i, a
16. e, g
17. h, f
18. c, j
19. d
20. d
21. a
22. b

Test Prep Pretest

1. c
2. a
3. a
4. d
5. d
6. c
7. c
8. gene
9. vector
10. complementary
11. insulin
12. factor VIII
13. pathogen
14. glyphosate
15. DNA ligase
16. The Human Genome Project is a research effort to identify, locate, and sequence the entire collection of genes in a human cell.
17. DNA fingerprints are used in paternity cases, forensics, and identifying genes that cause genetic disorders.
18. Because the human body contains small amounts of specific proteins that are required for proper functioning, large quantities of specific proteins needed to treat certain disorders are difficult to obtain. Genetic engineering technologies have allowed the production of large quantities of products containing the specific proteins required to treat many disorders.
19. If crops are naturally resistant to insects, farmers can use fewer pesticides, which are harmful to the environment.
20. Wilmut deprived the mammary cells (the differentiated cells) of nutrients before combining them with egg cells in which the nuclei had been removed. The nutrient deprivation forced the differentiated cells to pause at the beginning of the cell cycle so that both the differentiated cells and the egg cells would be at the same stage in their cycles at the moment they were combined.

Quiz

SECTION: GENETIC ENGINEERING

1. d
2. e
3. f
4. c
5. a
6. g
7. b
8. a
9. d
10. c

SECTION: HUMAN APPLICATIONS OF GENETIC ENGINEERING

1. f
2. e
3. a
4. b
5. c
6. d
7. d
8. b
9. a
10. c

SECTION: GENETIC ENGINEERING IN AGRICULTURE

1. d
2. c
3. a
4. b
5. e
6. d
7. a
8. d
9. c
10. b

Chapter Test (General)

1. f
2. i
3. j
4. e
5. b
6. k
7. g
8. h
9. d
10. a
11. c
12. b
13. d
14. c
15. b
16. d
17. c
18. a
19. b
20. d

Chapter Test (Advanced)

1. a
2. b
3. a
4. c
5. d
6. a
7. d
8. c
9. a
10. c
11. DNA fingerprint
12. plasmids
13. clones
14. restriction enzymes
15. Southern blot
16. probes
17. transgenic
18. differentiated
19. tetracycline
20. complementary, sticky
21. Answer should include any three of the following: making crops more tolerant to drought conditions; able to adapt to different soils, climates, and environmental stresses; resistant to the weedkiller glyphosate; resistant to insects; more nutritious
22. Answer should include any two of the following: to increase milk production by feeding cows GM growth hormone; increasing the weight of pigs by stimulating their natural growth hormone; and producing medically useful human proteins by adding human genes to those of livestock in order to get the animals to produce human proteins in their milk.
23. First, DNA from the organism containing the gene of interest and DNA from the vector, such as a plasmid, are cut with a restriction enzyme. Then the DNA fragments from the two sources are combined. The recombined DNA is inserted into bacteria. The recipient bacterial cells are cloned. Finally, the bacterial cells are screened to identify and isolate those specific bacterial cells that contain the transferred gene and that are producing the protein coded for by the gene.
24. DNA fingerprints are used in paternity cases, in forensics, and in the identification of the genes that cause genetic disorders.
25. Gel electrophoresis uses an electrical field within a gel to separate DNA fragments by their size and charge, allowing the fragments to be identified.

TEACHER RESOURCE PAGE

Lesson Plan

Section: Genetic Engineering

Pacing

Regular Schedule: with lab(s): 5 days   without lab(s): 3 days
Block Schedule: with lab(s): 2 1/2 days   without lab(s): 1 1/2 days

Objectives

1. Describe four basic steps commonly used in genetic engineering experiments.
2. Evaluate how restriction enzymes and the antibiotic tetracycline are used in genetic engineering.
3. Relate the role of electrophoresis and probes in identifying a specific gene.

National Science Education Standards Covered

UNIFYING CONCEPTS AND PROCESSES

UCP1: Systems, order, and organization

UCP2: Evidence, models, and explanation

SCIENCE AS INQUIRY

SI1: Abilities necessary to do scientific inquiry

SI2: Understandings about scientific inquiry

SCIENCE AND TECHNOLOGY

ST1: Abilities of technological design

ST2: Understandings about science and technology

SCIENCE IN PERSONAL AND SOCIAL PERSPECTIVES

SPSP6: Science and technology in local, national, and global challenges

HISTORY AND NATURE OF SCIENCE

HNS1: Science as a human endeavor

HNS2: Nature of scientific knowledge

HNS3: Historical perspectives

LIFE SCIENCE: THE CELL

LSCell3: Cells store and use information to guide their functions.

LSCell4: Cell functions are regulated.

Copyright © by Holt, Rinehart and Winston. All rights reserved.

TEACHER RESOURCE PAGE

Lesson Plan continued

LSCell6: Cells can differentiate, and complex multicellular organisms are formed as a highly organized arrangement of differentiated cells.

LIFE SCIENCE: MOLECULAR BASIS OF HEREDITY

LSGene1: In all organisms, the instructions for specifying the characteristics of the organisms are carried in DNA.

KEY
SE = Student Edition   TE = Teacher Edition
CRF = Chapter Resource File

Block 1
CHAPTER OPENER *(45 minutes)*

- **Quick Review,** SE. Students answer questions covered in previous sections of the textbook as preparation for the chapter content. **(GENERAL)**

- **Reading Activity,** SE. Students write a short list of things they know about gene technology and a list of things they want to know about gene technology. **(GENERAL)**

- **Using the Figure,** TE. Students answer questions about the chapter opener photograph. **(GENERAL)**

- **Opening Activity,** Gene Technology, TE. Students discuss why gene technology is controversial. **(GENERAL)**

Block 2
FOCUS *(5 minutes)*

- **Bellringer Transparency.** Use this transparency as students enter the classroom and find their seats. **(GENERAL)**

MOTIVATE *(10 minutes)*

- **Discussion,** TE. Discuss favorable traits in fruits and vegetables and link these traits to how scientists are able to manipulate many characteristics of fruits and vegetables using genetic engineering techniques. **(BASIC)**

TEACH *(30 minutes)*

- **Teaching Transparency, Section Outline.** Use this transparency to give students a framework for the information in this section. **(GENERAL)**

- **Teaching Transparency, Genetic Engineering.** Use this transparency to walk students through the four basic steps most genetic engineering experiments share. **(GENERAL)**

- **Activity,** Recombinant DNA, TE. Students use a piece of yarn, a long pipe cleaner, some tape, and scissors to represent a human chromosome and its components. **(GENERAL)**

Copyright © by Holt, Rinehart and Winston. All rights reserved.

TEACHER RESOURCE PAGE
Lesson Plan *continued*

- **Teaching Transparency, Restriction Enzymes Cut DNA.** Use this transparency to discuss how restriction enzymes work. (**GENERAL**)
- **Teaching Transparency, Southern Blot.** Use this transparency to walk students through the steps involved in the Southern blot technique. Explain that the Southern blot technique can be used for specific gene identification. (**GENERAL**)

HOMEWORK

- **Active Reading Worksheet, Genetic Engineering, CRF.** Students read a passage related to the section topic and answer questions. (**GENERAL**)
- **Directed Reading Worksheet, Genetic Engineering, CRF.** Students complete the exercises in this worksheet to help them understand the material as they read the section. (**BASIC**)

Block 3

TEACH *(30 minutes)*

- **Integrating Physics and Chemistry**, TE. Help students make the connection between electrophoresis and electrolytic behavior of solutions. (**GENERAL**)
- **Using the Figure**, Figure 5, TE. Use these tips to help students put all the pieces together in the figure. (**GENERAL**)
- **Quick Lab,** Modeling Gel Electrophoresis, SE. Students use beads to model how DNA fragments of different sizes are separated during gel electrophoresis. (**GENERAL**)
- **Datasheets for In-Text Labs,** Modeling Gel Electrophoresis, CRF.

CLOSE *(15 minutes)*

- **Reteaching**, TE. Students rewrite each objective as a question and then answer each question. (**BASIC**)
- **Quiz,** TE. Students answer questions that review the section material. (**GENERAL**)

HOMEWORK

- **Section Review,** SE. Assign questions 1–6 for review, homework, or quiz. (**GENERAL**)
- **Quiz, CRF.** This quiz consists of ten multiple choice and matching questions that review the section's main concepts. (**BASIC**) **Also in Spanish.**

Optional Blocks

LAB *(45 minutes plus 20 minutes on a second day)*

- **Skills Practice Lab, Transforming Bacteria with a Firefly Gene, CRF.** Students transform *E. coli* bacteria with the pBestLuc plasmid and determine if the transformation was sucessful by examining the cells for luminescence. (**GENERAL**)

Copyright © by Holt, Rinehart and Winston. All rights reserved.

Holt Biology / Gene Technology

TEACHER RESOURCE PAGE

Lesson Plan *continued*

Other Resource Options

- **Internet Connect.** Students can research Internet sources about Genetic Engineering with SciLinks Code HX4092.

- **go.hrw.com.** For worksheets, videos, and other teaching aids related to this chapter, visit the HRW Web site and type in the keyword HX4 GTC.

- **Biology Interactive Tutor CD-ROM,** Unit 6 Gene Expression. Students watch animations and other visuals as the tutor explains gene expression. Students assess their learning with interactive activities.

- **CNN Science in the News, Video Segment 8 Alzheimer's Mutation.** This video segment is accompanied by a **Critical Thinking Worksheet**.

- **CNN Student News.** Find the latest news, lesson plans, and activities related to important scientific events at **cnnstudentnews.com**.

TEACHER RESOURCE PAGE

Lesson Plan

Section: Human Applications of Genetic Engineering

Pacing

Regular Schedule: **with lab(s):** 5 days   **without lab(s):** 2 days
Block Schedule: **with lab(s):** 2 1/2 days   **without lab(s):** 1 day

Objectives

1. Summarize two major goals of the Human Genome Project.
2. Describe how drugs produced by genetic engineering are being used.
3. Summarize the steps involved in making genetically engineered vaccine.
4. Identify two different uses for DNA fingerprints.

National Science Education Standards Covered

UNIFYING CONCEPTS AND PROCESSES

UCP2: Evidence, models, and explanation

UCP5: Form and function

SCIENCE AS INQUIRY

SI1: Abilities necessary to do scientific inquiry

SI2: Understandings about scientific inquiry

SCIENCE AND TECHNOLOGY

ST1: Abilities of technological design

ST2: Understandings about science and technology

SCIENCE IN PERSONAL AND SOCIAL PERSPECTIVES

SPSP6: Science and technology in local, national, and global challenges

HISTORY AND NATURE OF SCIENCE

HNS1: Science as a human endeavor

HNS2: Nature of scientific knowledge

HNS3: Historical perspectives

LIFE SCIENCE: THE CELL

LSCell3: Cells store and use information to guide their functions.

LSCell4: Cell functions are regulated.

Copyright © by Holt, Rinehart and Winston. All rights reserved.

Holt Biology   Gene Technology

TEACHER RESOURCE PAGE

Lesson Plan *continued*

LIFE SCIENCE: MOLECULAR BASIS OF HEREDITY

LSGene1: In all organisms, the instructions for specifying the characteristics of the organisms are carried in DNA.

LSGene2: Most of the cells in a human contain two copies of each of 22 different chromosomes. In addition, there is a pair of chromosomes that determine sex.

LSGene3: Changes in DNA (mutations) occur spontaneously at low rates.

KEY
SE = Student Edition TE = Teacher Edition
CRF = Chapter Resource File

Block 4

FOCUS *(5 minutes)*

- **Bellringer Transparency.** Use this transparency as students enter the classroom and find their seats. **(GENERAL)**

MOTIVATE *(10 minutes)*

- **Demonstration**, TE. Use an ink pad and white paper to make a fingerprint of each student, and explain that everyone's DNA fingerprint is as individual as their standard fingerprint. **(BASIC)**

TEACH *(30 minutes)*

- **Teaching Transparency, Section Outline.** Use this transparency to give students a framework for the information in this section. **(GENERAL)**
- **Directed Reading Worksheet, Human Applications of Genetic Engineering, CRF.** Students complete the exercises in this worksheet to help them understand the material as they read the section. **(BASIC)**
- **Teaching Transparency, Genetically Engineered Medicine.** Use this transparency to point out examples of genetically engineered medicines available on the market. **(GENERAL)**
- **Group Activity**, Genetic Privacy, TE. Students form groups and discuss genetic information and privacy. Use the questions in this TE item to guide their discussions. **(GENERAL)**

HOMEWORK

- **Active Reading Worksheet, Human Applications of Genetic Engineering, CRF.** Students read a passage related to the section topic and answer questions. **(GENERAL)**

TEACHER RESOURCE PAGE

Lesson Plan *continued*

Block 5

TEACH *(30 minutes)*

- **Teaching Tip**, Graphic Organizer, TE. Students make a graphic organizer to outline how genetically engineered drugs are made. A sample graphic organizer is provided in the TE. (**GENERAL**)
- **Teaching Transparency, Genetically Engineered Vaccine.** Use this transparency to discuss how genetically engineered vaccines are made. (**GENERAL**)
- **Group Activity**, Smallpox, TE. Students work in small groups, each to research a different aspect of smallpox. (**ADVANCED**)
- **Real Life**, SE. Students find out the most common ways that vaccines are now administered. (**GENERAL**)

CLOSE *(15 minutes)*

- **Quiz,** TE. Students answer questions that review the section material. (**GENERAL**)
- **Reteaching,** TE. Students write one sentence for each of the key terms listed for this section. Each sentence should demonstrate the meaning of the term as it is defined in the text. (**BASIC**)

HOMEWORK

- **Alternative Assessment**, TE. Students prepare a list of questions about research conducted at a local university or pharmaceutical company. (**GENERAL**)
- **Quiz, CRF.** This quiz consists of ten multiple choice and matching questions that review the section's main concepts. (**BASIC**) **Also in Spanish.**
- **Section Review,** SE. Assign questions 1–6 for review, homework, or quiz. (**GENERAL**)

Optional Blocks

LAB *(135 minutes)*

- **Exploration Lab, DNA Fingerprinting, CRF.** Students perform some of the experimental procedures involved in DNA fingerprinting and use the results to identify a hypothetical burglar. (**GENERAL**)

Other Resource Options

- **Internet Connect.** Students can research Internet sources about Genetic Engineering with SciLinks Code HX4092.
- **go.hrw.com.** For worksheets, videos, and other teaching aids related to this chapter, visit the HRW Web site and type in the keyword HX4 GTC.

TEACHER RESOURCE PAGE

Lesson Plan continued

- **Biology Interactive Tutor CD-ROM,** Unit 6 Gene Expression. Students watch animations and other visuals as the tutor explains gene expression. Students assess their learning with interactive activities.
- **CNN Science in the News, Video Segment 8 Alzheimer's Mutation.** This video segment is accompanied by a **Critical Thinking Worksheet**.
- **CNN Student News.** Find the latest news, lesson plans, and activities related to important scientific events at **cnnstudentnews.com**.

TEACHER RESOURCE PAGE
Lesson Plan

Section: Genetic Engineering in Agriculture

Pacing

Regular Schedule: with lab(s): N/A without lab(s): 2 days
Block Schedule: with lab(s): N/A without lab(s): 1 day

Objectives

1. Describe three ways in which genetic engineering has been used to improve plants.
2. Summarize two ways in which genetic engineering techniques have been used to modify farm animals.
3. Summarize the cloning of sheep through the use of differentiated cells.

National Science Education Standards Covered

UNIFYING CONCEPTS AND PROCESSES

UCP1: Systems, order, and organization

UCP2: Evidence, models, and explanation

UCP3: Change, constancy, and measurement

UCP4: Evolution and equilibrium

UCP5: Form and function

SCIENCE AS INQUIRY

SI1: Abilities necessary to do scientific inquiry

SI2: Understandings about scientific inquiry

SCIENCE AND TECHNOLOGY

ST1: Abilities of technological design

ST2: Understandings about science and technology

SCIENCE IN PERSONAL AND SOCIAL PERSPECTIVES

SPSP6: Science and technology in local, national, and global challenges

HISTORY AND NATURE OF SCIENCE

HNS1: Science as a human endeavor

HNS2: Nature of scientific knowledge

HNS3: Historical perspectives

Copyright © by Holt, Rinehart and Winston. All rights reserved.

Holt Biology   Gene Technology

TEACHER RESOURCE PAGE

Lesson Plan *continued*

LIFE SCIENCE: THE CELL

LSCell3: Cells store and use information to guide their functions.

LSCell6: Cells can differentiate and form complete multicellular organisms.

LIFE SCIENCE: MOLECULAR BASIS OF HEREDITY

LSGene1: In all organisms, the instructions for specifying the characteristics of the organisms are carried in DNA.

LSGene3: Changes in DNA (mutations) occur spontaneously at low rates.

KEY
SE = Student Edition   TE = Teacher Edition
CRF = Chapter Resource File

Block 6

FOCUS *(5 minutes)*

- **Bellringer Transparency.** Use this transparency as students enter the classroom and find their seats. **(GENERAL)**

MOTIVATE *(10 minutes)*

- **Discussion**, TE. Lead a discussion on the first genetically altered fruit, Flavr-Saver™ tomatoes. Ask students what the advantages and disadvantages of altered crops might be. **(GENERAL)**

TEACH *(30 minutes)*

- **Teaching Transparency, Section Outline.** Use this transparency to give students a framework for the information in this section. **(GENERAL)**
- **Directed Reading Worksheet, Genetic Engineering in Agriculture, CRF.** Students complete the exercises in this worksheet to help them understand the material as they read the section. **(BASIC)**
- **Group Activity**, Genetically Engineered Crop Plants, TE. Students work in small groups to research genetically engineered crop or ornamental plants. **(GENERAL)**

HOMEWORK

- **Active Reading Worksheet, Genetic Engineering in Agriculture, CRF.** Students read a passage related to the section topic and answer questions. **(GENERAL)**

TEACHER RESOURCE PAGE

Lesson Plan *continued*

Block 7

TEACH *(30 minutes)*

- **Reading Skill Builder**, Brainstorming, TE. Students compare food crops and farm animals produced by selective breeding and genetic engineering. **(GENERAL)**
- **Group Activity**, Ads for Gene Products, TE. Students work in small groups to choose and then research a genetically engineered product such as human insulin. **(ADVANCED)**
- **Reading Skill Builder**, Discussion, TE. Students discuss the unforeseen problems that have developed in the cloned lamb, Dolly. Ask students why the chromosomes would appear abnormally old when Dolly is relatively young. **(ADVANCED)**

CLOSE *(15 minutes)*

- **Reteaching**, TE. Students create a table to summarize the ways in which genetic engineering has been used to improve food crops and farm animals and to make medically useful proteins in the milk of farm animals. **(BASIC)**
- **Quiz**, TE. Students answer questions that review the section material. **(GENERAL)**

HOMEWORK

- **Alternative Assessment**, TE. Students return to their lists of things they want to know about gene technology from the chapter opener. Have them check off the questions that they can now answer and make a list of what they have learned. Review any questions they still have. **(GENERAL)**
- **Section Review**, SE. Assign questions 1–5 for review, homework, or quiz. **(GENERAL)**
- **Science Skills Worksheet**, CRF. Students interpret diagrams showing the steps in a genetic engineering experiment. **(GENERAL)**
- **Quiz**, CRF. This quiz consists of ten multiple choice and matching questions that review the section's main concepts. **(BASIC) Also in Spanish.**
- **Modified Worksheet**, One-Stop Planner. This worksheet has been specially modified to reach struggling students. **(BASIC)**
- **Critical Thinking Worksheet**, CRF. Students answer analogy-based questions that review the section's main concepts and vocabulary. **(ADVANCED)**

Other Resource Options

- **Supplemental Reading, The Double Helix**, One-Stop Planner. Students read the book and answer questions. **(ADVANCED)**
- **Internet Connect.** Students can research Internet sources about Genetic Engineering with SciLinks Code HX4092.

Holt Biology — Gene Technology

TEACHER RESOURCE PAGE

Lesson Plan continued

- **go.hrw.com.** For worksheets, videos, and other teaching aids related to this chapter, visit the HRW Web site and type in the keyword HX4 GTC.
- **Biology Interactive Tutor CD-ROM,** Unit 6 Gene Expression. Students watch animations and other visuals as the tutor explains gene expression. Students assess their learning with interactive activities.
- **CNN Science in the News, Video Segment 8 Alzheimer's Mutation.** This video segment is accompanied by a **Critical Thinking Worksheet**.
- **CNN Student News.** Find the latest news, lesson plans, and activities related to important scientific events at **cnnstudentnews.com**.

TEACHER RESOURCE PAGE

Lesson Plan

End-of-Chapter Review and Assessment

Pacing
Regular Schedule: 2 days
Block Schedule: 1 day

> **KEY**
> **SE** = Student Edition   **TE** = Teacher Edition
> **CRF** = Chapter Resource File

Block 8
REVIEW *(45 minutes)*

- **Study Zone,** SE. Use the Study Zone to review the Key Concepts and Key Terms of the chapter and prepare students for the Performance Zone questions. **(GENERAL)**

- **Performance Zone,** SE. Assign questions to review the material for this chapter. Use the assignment guide to customize review for sections covered. **(GENERAL)**

- **Teaching Transparency, Concept Mapping.** Use this transparency to review the concept map for this chapter. **(GENERAL)**

Block 9
ASSESSMENT *(45 minutes)*

- **Chapter Test, Gene Technology, CRF.** This test contains 20 multiple choice and matching questions keyed to the chapter's objectives. **(GENERAL) Also in Spanish.**

- **Chapter Test, Gene Technology, CRF.** This test contains 25 questions of various formats, each keyed to the chapter's objectives. **(ADVANCED)**

- **Modified Chapter Test,** One-Stop Planner. This test has been specially modified to reach struggling students. **(BASIC)**

Other Resource Options

- **Vocabulary Review Worksheet, CRF.** Use this worksheet to review the chapter vocabulary. **(GENERAL) Also in Spanish.**

- **Test Prep Pretest, CRF.** Use this pretest to review the main content of the chapter. Each question is keyed to a section objective. **(GENERAL) Also in Spanish.**

- **Test Item Listing for ExamView® Test Generator, CRF.** Use the Test Item Listing to identify questions to use in a customized homework, quiz, or test.

- **ExamView® Test Generator, One-Stop Planner.** Create a customized homework, quiz, or test using the HRW Test Generator program.

Copyright © by Holt, Rinehart and Winston. All rights reserved.

Holt Biology   97   Gene Technology

TEST ITEM LISTING
Gene Technology

TRUE/FALSE

1. ____ Growing a large number of different cells from one cell is known as cloning.
   Answer: False   Difficulty: I   Section: 1   Objective: 1

2. ____ Manipulating genes for practical purposes is called genetic engineering.
   Answer: True   Difficulty: I   Section: 1   Objective: 1

3. ____ Gene cloning is an efficient way to produce many copies of a specific DNA sequence.
   Answer: True   Difficulty: I   Section: 1   Objective: 1

4. ____ Scientists have used genetic engineering to produce bacteria capable of synthesizing human proteins.
   Answer: True   Difficulty: I   Section: 1   Objective: 1

5. ____ Gene cloning is an efficient means of producing large numbers of different genes.
   Answer: False   Difficulty: I   Section: 1   Objective: 1

6. ____ In bacteria, a circular DNA molecule that replicates independently of the main chromosome is called a plasmid.
   Answer: True   Difficulty: I   Section: 1   Objective: 1

7. ____ In the practice of genetic engineering, scientists directly manipulate genes.
   Answer: True   Difficulty: I   Section: 1   Objective: 1

8. ____ Plasmids are pieces of viral DNA that commonly infect human cells.
   Answer: False   Difficulty: I   Section: 1   Objective: 1

9. ____ DNA ligase can seal the sticky ends of a DNA fragment.
   Answer: True   Difficulty: I   Section: 1   Objective: 1

10. ____ Before a foreign gene is inserted into a plasmid, the plasmid is opened with a restriction enzyme.
    Answer: True   Difficulty: I   Section: 1   Objective: 2

11. ____ Recombinant DNA is made when a DNA fragment is put into the DNA of a vector.
    Answer: True   Difficulty: I   Section: 1   Objective: 2

12. ____ Restriction enzymes make a straight cut through both strands of DNA.
    Answer: False   Difficulty: I   Section: 1   Objective: 2

13. ____ Gel electrophoresis separates DNA fragments by their size and shape.
    Answer: True   Difficulty: I   Section: 1   Objective: 3

14. ____ In a Southern blot, the DNA from each bacterial colony is isolated and cut into fragments by probes.
    Answer: False   Difficulty: I   Section: 1   Objective: 3

15. ____ The effort to catalog, locate, and sequence all the chromosomes of every living organism is a goal of the Human Genome Project.
    Answer: False   Difficulty: I   Section: 2   Objective: 1

16. ____ Factor VIII is a protein that promotes blood clotting.
    Answer: True   Difficulty: I   Section: 2   Objective: 2

17. ____ Injection of a particular vaccine can cause the body to produce antibodies that protect against the possibility of future infection by a particular pathogen.
    Answer: True   Difficulty: I   Section: 2   Objective: 3

TEST ITEM LISTING, continued

18. ____ DNA fingerprinting enables genetic engineers to arrange genes in a particular order on a chromosome.
    Answer: False   Difficulty: I   Section: 2   Objective: 4

19. ____ DNA fingerprint analysis can be used to determine whether two individuals are related.
    Answer: True   Difficulty: I   Section: 2   Objective: 4

20. ____ RFLPs are pieces of DNA that are all the same length.
    Answer: True   Difficulty: I   Section: 2   Objective: 4

21. ____ If a crop is made glyphosate-resistant, treating it with glyphosate will seriously reduce its yield.
    Answer: False   Difficulty: I   Section: 3   Objective: 1

22. ____ Despite the potential environmental benefits, genetic engineers have been unable to develop crop plants that are resistant to weedkillers.
    Answer: False   Difficulty: I   Section: 3   Objective: 1

23. ____ Genetic engineering techniques can be used to make crops resistant to destructive insects.
    Answer: True   Difficulty: I   Section: 3   Objective: 1

24. ____ Genetic engineers have developed a method of infecting cows with milk-producing bacteria to increase the amount of milk produced by the cows.
    Answer: False   Difficulty: I   Section: 3   Objective: 2

25. ____ Dairy cattle will produce more milk when genetically engineered growth hormone is added to their food.
    Answer: True   Difficulty: I   Section: 3   Objective: 2

26. ____ A transgenic animal is an animal with foreign DNA in its cells.
    Answer: True   Difficulty: I   Section: 3   Objective: 2

MULTIPLE CHOICE

27. A strand of DNA formed by the splicing of DNA from two different species is called
    a. determinant RNA.
    b. recombinant DNA.
    c. plasmid DNA.
    d. restriction RNA.
    Answer: B   Difficulty: I   Section: 1   Objective: 1

28. Which of the following procedures is *not* a usual step in a genetic engineering experiment?
    a. inducing a mutation in a source chromosome
    b. cleaving DNA with a restriction enzyme
    c. recombining pieces of DNA from different species
    d. cloning and screening target cells
    Answer: A   Difficulty: I   Section: 1   Objective: 1

29. Genetic engineering refers to the process of
    a. creating new DNA molecules from nucleotide sequences.
    b. rearranging nucleotides in a gene of an organism so that new traits appear in the development of an embryo.
    c. moving genes from a chromosome of one organism to a chromosome of a different organism.
    d. building a new species by combining genes of different organisms.
    Answer: C   Difficulty: I   Section: 1   Objective: 1

TEST ITEM LISTING, continued

30. Cohen and Boyer transferred a gene from a frog chromosome into the genetic material of a
    a. different frog.
    b. different chromosome of the same frog.
    c. virus taken from the same frog.
    d. bacterial cell.

    Answer: D   Difficulty: I   Section: 1   Objective: 1

31. The use of genetic engineering to transfer human genes into bacteria
    a. is impossible with current technology.
    b. causes the human genes to manufacture bacterial proteins.
    c. results in the formation of a new species of organism.
    d. allows the bacteria to produce human proteins.

    Answer: D   Difficulty: I   Section: 1   Objective: 1

32. Cloning is a process by which
    a. undesirable genes may be eliminated.
    b. many identical protein fragments are produced.
    c. a virus and a bacterium may be fused into one.
    d. many identical cells may be produced.

    Answer: D   Difficulty: I   Section: 1   Objective: 1

33. Plasmids
    a. are circular pieces of bacterial DNA.
    b. can replicate independently of the organism's main chromosome.
    c. are often used as vectors in genetic engineering experiments.
    d. All of the above

    Answer: D   Difficulty: I   Section: 1   Objective: 1

34. Recombinant DNA is formed by joining DNA molecules
    a. from two different species.
    b. from two chromosomes of the same organism.
    c. with RNA molecules.
    d. with proteins from a different species.

    Answer: A   Difficulty: I   Section: 1   Objective: 1

35. plasmid : DNA segment coding for an enzyme ::
    a. DNA ligase : double-stranded DNA
    b. vector : restriction enzyme
    c. cloned cell : DNA ligase
    d. recombinant DNA : DNA from another organism

    Answer: D   Difficulty: II   Section: 1   Objective: 1

36. Restriction enzymes are specific in their action on
    a. DNA.   c. proteins.
    b. amino acids.   d. chromosomes.

    Answer: A   Difficulty: I   Section: 1   Objective: 2

37. Enzymes that cut DNA molecules at specific places
    a. have sticky ends.
    b. are restriction enzymes.
    c. work only on bacterial DNA.
    d. always break the DNA between guanine and adenine.

    Answer: B   Difficulty: I   Section: 1   Objective: 2

TEST ITEM LISTING, continued

38. After cloning bacteria that had been exposed to the recombinant DNA, Cohen and Boyer added tetracycline to the culture in order to
    a. kill any contaminating viruses.
    b. kill cells that did not have the recombinant DNA in their genomes.
    c. neutralize any frog genes that might remain.
    d. make the bacterial cells multiply faster.
    Answer: B   Difficulty: I   Section: 1   Objective: 2

39. DNA fragments cut by a restriction enzyme can
    a. pair up and join with any other DNA fragments cut by the same restriction enzyme.
    b. pair only with fragments formed by a complementary restriction enzyme.
    c. combine with any other spliced chromosome.
    d. pair only with DNA from the same species.
    Answer: A   Difficulty: II   Section: 1   Objective: 2

40. Radioactive or fluorescent-labeled RNA or single-stranded DNA pieces that are complementary to the gene of interest and are used to confirm the presence of a cloned gene are called
    a. probes.   c. vaccines.
    b. plasmids.   d. clones.
    Answer: A   Difficulty: I   Section: 1   Objective: 3

41. A technique that uses radioactively labeled DNA to identify specific genes in a piece of DNA is called the
    a. Northern blot.   c. Northern lights.
    b. Southern vector.   d. Southern blot.
    Answer: D   Difficulty: I   Section: 1   Objective: 3

42. The goal of the Human Genome Project is to
    a. create maps showing where genes are located on human chromosomes.
    b. create maps showing where chromosomes are located on human genes.
    c. treat patients with genetic diseases.
    d. identify people with genetic diseases.
    Answer: A   Difficulty: I   Section: 2   Objective: 1

43. A medical condition that can be treated by using proteins produced through genetic engineering is
    a. diabetes.
    b. ovarian cancer.
    c. hemophilia.
    d. All of the above
    Answer: D   Difficulty: I   Section: 2   Objective: 2

44. genetic engineering : human health ::
    a. vaccine : anticoagulant   c. anticoagulant : hemophilia
    b. human insulin : diabetes patients d. diabetes : insulin
    Answer: B   Difficulty: II   Section: 2   Objective: 2

45. factor VIII : hemophilia ::
    a. factor VIII : diabetes   c. blood factors : viruses
    b. growth hormone : diabetes   d. insulin : diabetes
    Answer: D   Difficulty: II   Section: 2   Objective: 2

TEST ITEM LISTING, continued

46. Antibodies
    a. prevent diseases caused by vaccines.
    b. are produced by bacteria that infect animals.
    c. help destroy microbes that invade the body.
    d. cause viruses to infect bacterial cells.
    Answer: C   Difficulty: I   Section: 2   Objective: 3

47. The risk associated with vaccines prepared by injecting killed or weakened pathogenic microbes is that
    a. a few remaining live or unweakened microbes could still cause the disease.
    b. the antibodies that result may not work.
    c. the vaccine protects only against other diseases.
    d. None of the above
    Answer: A   Difficulty: I   Section: 2   Objective: 3

48. DNA fingerprinting has been used in criminal investigations because
    a. criminals leave DNA samples behind them when they touch an object at a crime scene.
    b. DNA analysis allows investigators to distinguish body cells of different individuals, who are unlikely to have the same DNA.
    c. bacterial DNA on the hands of criminals may provide a clue as to where that person was when the crime was committed.
    d. DNA found on murder weapons is easy to identify.
    Answer: B   Difficulty: II   Section: 2   Objective: 4

49. A gene that codes for resistance to glyphosate has been added to the genome of certain plants. These plants will
    a. produce chemicals that kill weeds growing near them.
    b. die when exposed to glyphosate.
    c. convert glyphosate to fertilizer.
    d. survive when glyphosate is sprayed on the field.
    Answer: D   Difficulty: I   Section: 3   Objective: 1

50. Which of the following is *not* an example of gene technology used in farming?
    a. the use of cow growth hormone produced by bacteria to increase milk production in cows
    b. the development of larger and faster-growing breeds of livestock
    c. the cloning of human brain cells from selected farm animals
    d. the addition of human genes to farm-animal genes to obtain milk containing human proteins
    Answer: C   Difficulty: I   Section: 3   Objective: 2

51. Ian Wilmut's cloning of Dolly in 1997 was considered a breakthrough in genetic engineering because
    a. scientists thought cloning was impossible.
    b. scientists thought only fetal cells could be used to produce clones.
    c. scientists had never before isolated mammary cells.
    d. sheep had never responded well to gene technology procedures.
    Answer: B   Difficulty: I   Section: 3   Objective: 3

COMPLETION

52. The process by which a foreign gene is replicated by insertion into a bacterium is called _____ _____.
    Answer: gene cloning   Difficulty: II   Section: 1   Objective: 1

TEST ITEM LISTING, continued

53. A(n) _____ is an agent that is used to carry a DNA fragment isolated from one cell into another cell.
    Answer: vector   Difficulty: II   Section: 1   Objective: 1

54. Small, circular forms of bacterial DNA molecules that can replicate independently of the main bacterial chromosomes are called _____.
    Answer: plasmids   Difficulty: II   Section: 1   Objective: 1

55. Splicing DNA from two different organisms produces a new DNA segment called _____ _____.
    Answer: recombinant DNA   Difficulty: II   Section: 1   Objective: 1

56. A large number of genetically identical cells grown from a single cell are called _____.
    Answer: clones   Difficulty: I   Section: 1   Objective: 1

57. The process of allowing cells to reproduce in order to obtain a large number of identical cells is called _____.
    Answer: cloning   Difficulty: I   Section: 1   Objective: 1

58. Bacterial enzymes that cut DNA segments into shorter pieces are called _____ _____.
    Answer: restriction enzymes   Difficulty: II   Section: 1   Objective: 2

59. Enzymes that cleave DNA at specific sequences, generating a set of small fragments of DNA, are called _____ _____.
    Answer: restriction enzymes   Difficulty: II   Section: 1   Objective: 2

60. A technique known as _____ _____ can be used to separate molecules in a mixture by subjecting them to an electrical field within a gel.
    Answer: gel electrophoresis   Difficulty: II   Section: 1   Objective: 3

61. The name of the scientific program that has worked to construct maps of human chromosomes and determine the DNA sequences of those chromosomes is the _____ _____ _____.
    Answer: Human Genome Project   Difficulty: I   Section: 2   Objective: 1

62. The entire collection of genes within the cells of a human is referred to as the _____ _____.
    Answer: human genome   Difficulty: II   Section: 2   Objective: 1

63. The protein _____ is produced by genetic engineering to treat diabetes.
    Answer: insulin   Difficulty: II   Section: 2   Objective: 2

64. Defensive proteins that combat specific pathogens and stop their growth before they can cause disease are called _____.
    Answer: antibodies   Difficulty: I   Section: 2   Objective: 3

65. Crop plants have recently been developed that are resistant to the chemical _____, a powerful weedkiller.
    Answer: glyphosate   Difficulty: II   Section: 3   Objective: 1

66. A(n) _____ _____ was used to fuse mammary cells from one sheep with egg cells without nuclei from a different sheep.
    Answer: electric shock   Difficulty: II   Section: 3   Objective: 3

TEST ITEM LISTING, continued

ESSAY

67. Describe how a human gene may be recombined with an *E. coli* plasmid.
    Answer:
    First the plasmid is removed from the *E. coli* and opened with a restriction enzyme. The human gene is cut with the same restriction enzyme. Then the cut human gene is mixed with the cut plasmid. The sticky ends of the gene and plasmid join, resulting in recombinant DNA.
    Difficulty: III   Section: 1   Objective: 1

68. A scientist has produced a bacterium containing a human gene that codes for a useful protein. How can the scientist use gene cloning to produce large quantities of this protein?
    Answer:
    First the scientist should place the bacterium in culture medium to allow the cell to divide. Each new cell produced through division will contain the human gene. As these genes are expressed, the protein will be produced and can be harvested from the culture.
    Difficulty: III   Section: 1   Objective: 1

69. In order to transfer a gene from a member of one species to another, four distinct steps must be followed. Identify, in the correct order, the four steps of a genetic engineering experiment.
    Answer:
    1. The DNA from the source organism and the DNA from the vector are cleaved into fragments with the same restriction enzyme.
    2. Recombinant DNA is then produced by combining the cut source DNA fragment with the cut vector DNA.
    3. The recombined vector DNA is inserted into bacterial cells. The gene is cloned when bacteria are allowed to reproduce.
    4. The cloned target cells are screened to select those cells that contain the desired gene.
    Difficulty: III   Section: 1   Objective: 1

70. A scientist has a long segment of sequenced DNA that contains a gene that he would like to clone. However, the segment of DNA containing the gene is too large to insert into a bacterial plasmid. How might the scientist reduce the size of the fragment containing the gene?
    Answer:
    The scientist could use restriction enzymes to cut the DNA into smaller pieces. By choosing the right restriction enzymes, the scientist could create a smaller segment that still contained the gene to be cloned. This smaller fragment could then be combined with the bacterial plasmid.
    Difficulty: III   Section: 1   Objective: 2

71. Genetic engineering has made it possible for pharmaceutical companies to produce medicines such as insulin and human growth hormone. Give at least two reasons why this is important.
    Answer:
    The production of genetically engineered medicines is important because
    a. large amounts of medicines can be produced.
    b. it is safer—the process eliminates risk of contamination from diseases such as AIDS in human blood products.
    c. it is less expensive.
    Difficulty: III   Section: 2   Objective: 2

TEST ITEM LISTING, *continued*

72. Explain how a harmless virus might be turned into a vaccine by using genetic engineering.

    Answer:
    A DNA fragment coding for a surface protein of a disease-causing organism is inserted into the genome of a harmless virus. The recombinant virus is allowed to infect the organism that is to be protected. The recipient organism's body will respond by making antibodies that attack the surface protein of the disease-causing organism. If the vaccinated organism is ever exposed to the actual disease-causing organism, the vaccinated organism will immediately produce large amounts of the desired antibody to defend itself.

    Difficulty: III   Section: 2   Objective: 3

73. One of the greatest benefits of genetic engineering has been the manipulation of genes in crop plants such as wheat and soybeans. In what ways can genetic engineering affect agriculture?

    Answer:
    Crop plants can be genetically engineered to add favorable characteristics, including improved yields and resistance to weedkillers and destructive pests. Genetically engineered growth hormone increases milk production in dairy cows and weight gain in cattle and hogs. Transgenic animals can be cloned and used to make human proteins that are useful in medicine.

    Difficulty: III   Section: 3   Objective: 2

74. Summarize the cloning of sheep through the use of differentiated cells.

    Answer:
    Cells from the udder of one sheep were isolated and grown in nutrient-deficient solution. This stopped the cell cycle of the cells. Egg cells from another sheep were extracted, and their nuclei were removed. An electric shock was then used to open the egg cell, fuse the udder cell and the egg cell, and trigger the cell cycle to begin again. The embryo was later implanted into a surrogate mother sheep.

    Difficulty: III   Section: 3   Objective: 3